BUILDING SKILLS

Activity Lab Workbook

Contents

LIFE SCIENCE

Chapter 1 Plants and Animals 4
Chapter 2 Animals 21
Chapter 3 Looking at Habitats 37
Chapter 4 Kinds of Habitats 53

EARTH SCIENCE

Chapter 5 Land and Water 67
Chapter 6 Earth's Resources 81
Chapter 7 Observing Weather 97
Chapter 8 Earth and Space 111

PHYSICAL SCIENCE

Chapter 9 Looking at Matter 131
Chapter 10 Changes in Matter 145
Chapter 11 How Things Move 161
Chapter 12 Using Energy 181

Dear Parent or Guardian,

Today our science class talked about how to work safely when doing laboratory experiments. It is important that you be informed regarding the school's effort to promote a safe environment for students participating in laboratory activities. Please review the safety rules and this entire Safety Contract with your child. This contract must be signed by both you and your child in order for your child to participate in laboratory activities.

Safety Rules:

1. Listen carefully and follow directions.
2. If you are not sure about doing something, ask your teacher.
3. Never run or throw anything unless instructed differently by the activity.
4. Never taste anything when doing a science activity.
5. Always wash your hands before and after an activity.
6. Cooperate with others when working in a group.
7. Always clean up when you have finished.

Date: ___________

I have read and reviewed the science safety rules with my child. I consent to my child's participation in science laboratory activities in a classroom environment where these rules are enforced.

Parent/Guardian signature: ______________________

I know that it is important to work safely in science class. I understand the rules and will follow them.

Student signature: ______________________

Acuerdo de Seguridad para Ciencias

Estimados padres o tutor:

Hoy hemos hablado en nuestra clase de Ciencias sobre cómo mantener la seguridad al realizar experimentos científicos. Es importante que ustedes estén informados del propósito de la escuela de promover un entorno seguro para los estudiantes que participan en las prácticas de laboratorio. Por favor, examinen cuidadosamente con su niño o niña las reglas siguientes y el Acuerdo de Seguridad. El acuerdo debe ser firmado tanto por uno de ustedes como por su niño o niña para que él o ella pueda participar en las actividades de laboratorio.

Reglas de Seguridad:

1. Escucha con atención y sigue las indicaciones.
2. Si no estás seguro de algo pregúntale a tu maestro o maestra.
3. No corras ni arrojes ningún objeto a menos que sea parte de una actividad.
4. No te lleves nada a la boca ni lo pruebes cuando estés realizando una actividad de ciencias.
5. Lávate siempre las manos antes y después de una actividad.
6. Coopera con tus compañeros cuando estés trabajando en grupo.
7. No te olvides de limpiar cuando hayas terminado una actividad.

Fecha: ____________

He leído y examinado las reglas de seguridad de ciencias con mi niño o niña. Doy mi consentimiento para su participación en las actividades del laboratorio de ciencias en un entorno donde se hagan cumplir estas reglas.

Firma de uno de los padres o tutor: ____________________________

Sé la importancia que tiene trabajar con seguridad en la clase de Ciencias. Comprendo las reglas y me comprometo a seguirlas.

Firma del estudiante: _________________________________

Name ______________________ Date __________

Explore

How can a frog float on a lily pad?

You need

- paper plate
- green crayon
- scissors
- string
- pan of water
- toy frog

What to Do

1. **Predict.** Where should you place the frog on the lily pad so that the frog stays dry?

____

____

____

____

2. **Make a Model.** Color a paper plate green with a crayon. This will be the lily pad.

3. △ **Be Careful.** Use scissors to poke a small hole near the edge of the lily pad. Tie a six-inch piece of string through the hole.

4. Place the lily pad in a pan of water with the string below it.

 Name ______________________ Date ____________

5 **Record Data.** Draw and write down where you placed the frog.

____

____

____

Name ______________________ Date __________

What can carry clay on water?

You need

- classroom objects
- ball of clay
- pan of water

In this activity, you will find something that will float while carrying a small ball of clay.

What to Do

1. **Predict.** What do you think you can use to carry a ball of clay on top of the water?

____

____

2. Look around the classroom and find an object that you think will float in a pan of water.
3. Put the object you chose into a pan of water, and place a ball of clay on top of it. Does the object float or sink?

____

What Did You Find Out?

4. Why do you think your object was or was not able to float?

____

Explore Name ____________________ Date __________

How does a frog move?

What to Do

1. **Observe.** Look at the pictures on this page. Think about how the frogs are moving.

2. **Record Data.** Make a list of the different ways you see the frogs moving.

____

____

____

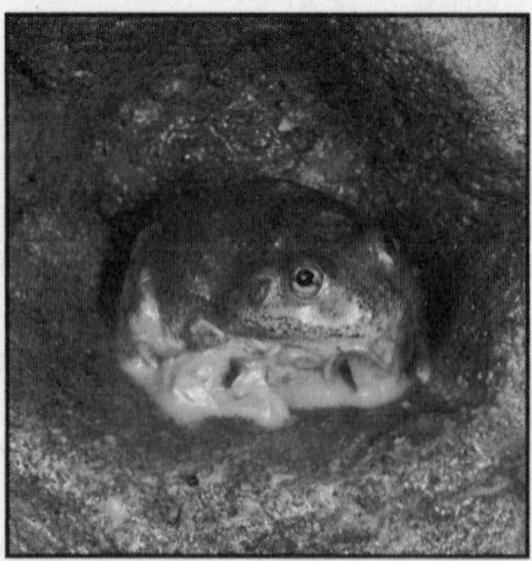

Name ______________________ Date ____________

Explore

3 Draw Conclusions. Add to your list. Write the body part the frogs use to move in each way.

____

____

____

____

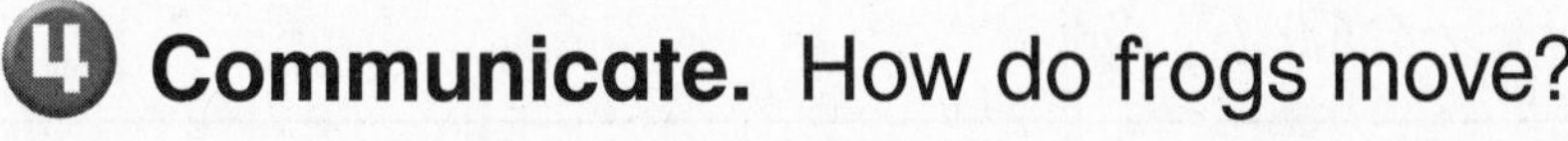

4 Communicate. How do frogs move?

____

____

____

Alternative Explore

Name ______________________ Date __________

What lives in or near a pond?

In this activity, you will find out what lives in or near ponds.

You need

- photos of ponds

What to Do

1. **Observe.** Look carefully at a picture of a pond.
2. **Record Data.** Write down what you observe about what lives in or near ponds.

____

____

3. **Communicate.** Discuss with a partner what you observed and the conclusions you drew about what lives in or near a pond.

What Did You Find Out?

4. How did recording what you saw in the pictures help you communicate your conclusions?

____

____

____

Name ______________________ Date __________

Explore

What do leaves need?

What to Do

1. Put the plants in a sunny place. Choose one plant and cover its leaves with foil. Keep the soil moist in both pots.

2. **Predict.** What will happen to each plant in a week?

____

____

____

 Name ________________ Date __________

3 **Record Data** Write down what you observe for a week.

4 Were your predictions correct? What do leaves need?

____

____

____

Explore More

5 **Predict.** What will happen if the foil is removed? Observe the plant for a week. Was your prediction correct?

____

____

____

Name ______________________ Date __________

How do leaves help a plant get light?

You need

- pictures of plants

In this activity, you will observe how the shape of leaves helps a plant get sunlight.

What to Do

1. **Compare.** Look at the pictures and compare the leaves of the plants. How are they alike? How are they different?

   ____

   ____

2. Hold out your hand in a sunny spot. Open and close your hand. Turn it from side to side. How do you need to hold your hand to get the most light?

   ____

   ____

3. **Compare** the shape of a leaf to the shape of your hand when you hold it out flat. How do the shapes of leaves help plants get sunlight?

   ____

Quick Lab

Name ____________________ Date __________

Observe Two Plants

You need

- 2 small potted plants
- water
- sponge
- hand lens
- paper
- crayons

What to Do

1. Observe how plants take in water. Do you think the roots or the leaves soak up water?

____

____

____

____

2. Gently take the plants out of the soil.

3. Use the hand lens to observe the roots. Draw a picture of what the roots look like through the hand lens.

4. Compare plants A and B. What happened? Why?

____

____

____

Name ______________________ Date ____________

Focus on Skills

Observe

Learn It

To **observe**, you use your senses to learn about something. You use senses to see, hear, taste, smell, and touch.

Learn It

You can use some of your senses to learn about flowers. You can make a chart to write down what you observe.

jasmine

see	
feel	The leaves feel smooth.
hear	
smell	The flowers smell sweet.

Name ______________________ Date __________

Try It

Find a flower to observe or look at the pictures below.

1. What color is your flower? Which sense did you use to find out?

____

____

2. How do you think the leaves will feel to your touch?

____

3. **Write About It.** Find another flower and compare.

____

____

bougainvillea

yucca plant

Name ______________________ Date ____________

Explore

What are the parts of a seed?

What to Do

You need

- wet lima bean 
- dry lima bean 
- hand lens 

1. **Observe.** What does the outside of a dry lima bean feel like? Use a hand lens. What do you see?

____

____

____

2. **Predict.** Draw what you think is inside the seed.

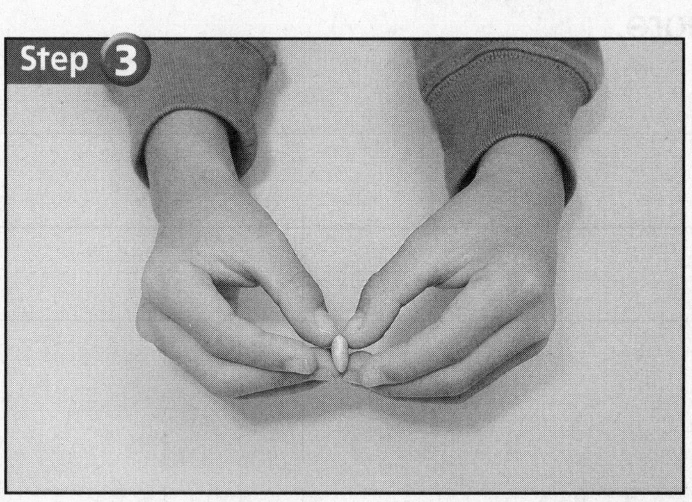

 Name ______________ Date ________

3 Use your fingernail to open the wet seed. Use your hand lens to observe the wet seed. Draw what you see.

4 **Communicate.** Compare your two drawings. What was different? What was the same?

____

____

____

Explore More

5 **Observe.** Look at other wet and dry seeds to see how they compare.

____

____

____

Name ______________________ Date __________

What are the parts of seeds?

In this activity, you will label the parts of a seed.

What to Do

1. Look at the diagram of the inside and outside of a bean seed.

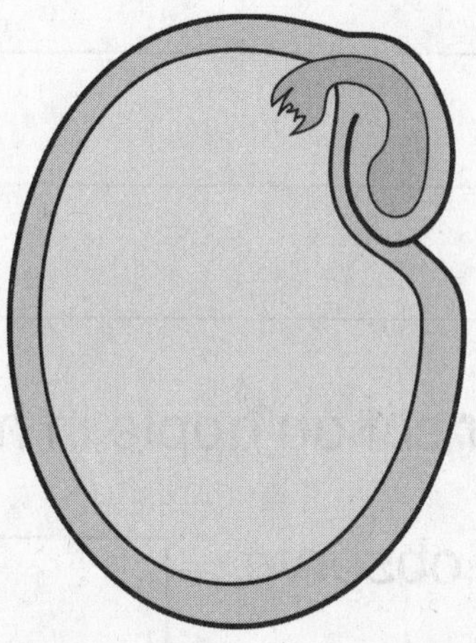

2. Label the seed coat.
3. Label the area of the seed where food is stored.
4. Label the part of the seed where a new plant will grow.

Quick Lab Name ______________________ Date __________

Observe Apple Seeds

You need

- apple
- knife
- crayons
- paper

What to Do

1. **Observe.** Where are the seeds found in the apple? Why do you think they are found there?

____

____

____

2. Watch your teacher cut an apple in half.

3. Use a hand lens to observe the seeds. Draw what you see.

4. Discuss with classmates what you know about seeds. Why are seeds important?

____

____

Name ______________________ Date __________

How do roots grow?

What to Do

1 Put three bean seeds on a damp paper towel. Put them in the bag and tape it to a bulletin board.

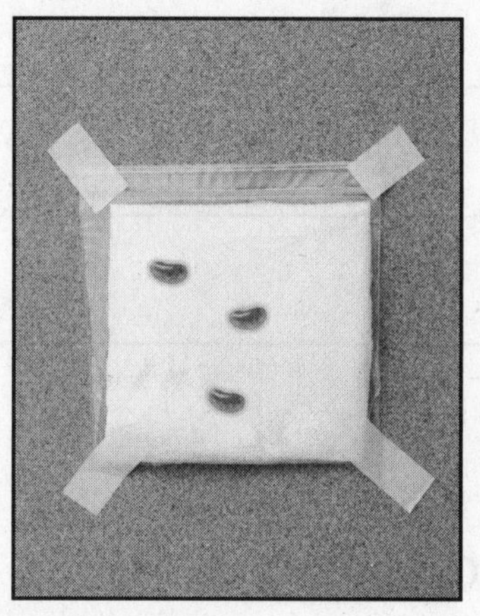

You need

- 3 bean seeds 
- paper towels 
- tape 
- plastic bag 
- hand lens 

2 **Observe.** Which part grows first? Which way did the roots grow?

____

____

____

 Name ________________ Date __________

3. After the roots grow, turn the bag upside down. Tape it to the board again. Make sure the paper towel stays wet.

4. **Draw Conclusions.** What happened to the roots?

____

____

____

Explore More

5. **Investigate.** What happens to the roots if left in the dark?

____

____

____

Name ______________________ Date __________

Alternative Explore

What makes seeds sprout down?

You need

- 4 bean seeds
- plastic bag
- moist paper towel

In this activity, you will observe and compare plants.

What to Do

1. Observe bean seed roots.
2. Place 4 beans in a plastic bag. Turn the beans in different directions. Put a moist paper towel in the bag with the beans.
3. **Observe** the bean seeds every day until the roots begin to grow.

What Did You Find Out?

4. How did the roots of each seed grow?

____

____

5. How do the roots know which way is down?

____

6. Why do you think roots grow down?

____

____

Quick Lab

Name ____________________ Date __________

Plants and Light

What to Do

You need

- 2 small potted seedlings
- water
- shoebox with hole
- crayons

1. Observe two seedlings to find out if plants grow toward light.
2. Place one of the seedlings in the shoebox. The shoebox should have a hole on one side. Put both plants next to the window for a week. Only open the box to water the seedling.
3. Open the box. What happened?

____

____

____

4. Imagine that you left the boxed seedling near the window for a month. Draw what you think it would look like when you opened the shoebox.

Name ______________________ Date __________

How can we put animals into groups?

What to Do

1. **Classify.** Look at the pictures of the animals. Put the animals into groups. How did you decide to group the animals?

____

____

____

2. Talk about the animal groups with a partner. What groups did your partner use?

____

____

____

 Name ____________ Date ________

Explore More

3 **Compare.** How are your groups and your partner's groups alike? How are they different?

____

____

____

4 **Classify.** Think about animals that live on land. How can you classify them?

____

____

____

Name ______________________ Date __________

How are animals alike and different?

You need

- pictures of animals
- pencil

In this activity, you will explore the ways that animals are alike and different.

What to Do

1. **Observe.** Look carefully at a picture of an animal and describe the animal to your partner.

2. **Compare.** Discuss how your animals are the same and different. Fill in the Venn diagram with the similarities and differences between the animals you and your partner chose.

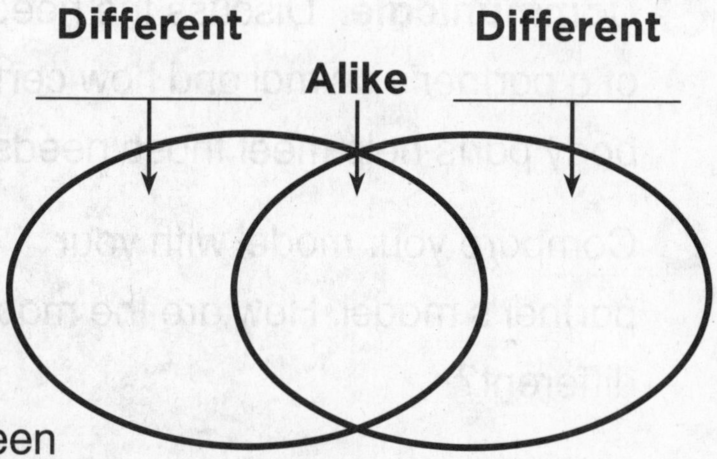

Different ______ Alike ______ Different ______

What Did You Find Out?

3. How do different animals meet their needs?

____

____

____

____

Quick Lab

Name ______________________ Date __________

Make an Animal Model

What to Do

You need
• heavy paper
• colored paper
• shiny paper
• felt
• foil
• chenille sticks
• glue
• scissors
• tape

1. Choose an animal and use the art materials to make a model of your animal. Use the art materials to show important body parts that help animals meet their needs.

2. **Communicate.** Discuss the needs of a partner's animal and how certain body parts help meet those needs.

3. Compare your model with your partner's model. How are the models alike and different?

   ____

4. **Discuss.** How do your animals use their bodies to meet their needs?

   ____

   ____

   ____

Name ______________________ Date __________

Focus on Skills

Classify

When you classify, you put things into groups to show how they are alike.

Learn It

You can use a chart to classify what you learned about animals.

Different Animals		
Mammals	Reptiles	Birds
sheep	lizard	eagle
mouse	turtle	sparrow

Try It

Use a chart to classify these animals. Add other animals to your chart. Share your chart with a partner.

Focus on Skills

Name ______________________ Date ____________

1. How are mammals and birds alike? How are they different?

____

____

____

2. What ways did you classify the animals in your chart?

____

____

____

3. **Write About It.** How is your chart different from the chart your partner made?

____

____

____

Name ______________________ Date __________

Explore

How are babies and adults alike and different?

What to Do

1. What are some things that babies do?

____

____

____

2. What are some things adults do?

____

____

____

Name ______________________ Date ____________

Explore More

3 **Compare.** Make a Venn diagram to compare babies to adults.

4 How are baby humans and baby tigers alike and different?

____

____

____

Name ______________________ Date __________

How can you compare baby and adult animals?

In this activity you will observe and compare baby and adult animals.

You need

- pictures of adult animals
- pictures of baby animals

What to Do

1. **Classify.** Sort your pictures into two groups. You should have an adult group and a baby group.
2. **Observe.** Work with a partner to carefully observe what the baby animals look like and what they are doing in the pictures.
3. **Observe.** Look carefully at the pictures of the adult animals and observe what they look like and what they are doing in the pictures.

What Did You Find Out?

4. How are some baby animals and adult animals the same? How are they different?

____

____

____

Quick Lab Name ______________________ Date __________

Act Out an Animal Life Cycle

You need

- various materials for children to use for skits

What to Do

1. Work with your group to decide which animal's life cycle you want to act out. Look in Chapter 2, Lesson 2 of your book for ideas if you are unsure of how the life cycle works.

2. **Communicate.** Study the animal you want to act out so that you can communicate and act out the life cycle to others.

3. The first step in a chicken's life cycle is an egg. How can you act like an egg? Try it out!

____

____

4. How are the life cycles of animals the same and different?

____

____

____

____

Name ______________________ Date ____________

Be a Scientist

How does a mealworm grow?

Find out how a mealworm grows and changes.

You need

- oatmeal
- container with lid 
- hand lens 
- mealworm larva 
- slice of apple 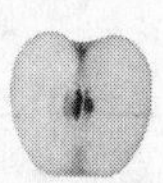
- ruler

What to Do

1. Put some oatmeal in the container. Poke holes in the lid.

2. **Observe.** What does a mealworm look like? Use your hand lens to observe the mealworm. Place a mealworm and an apple slice in the container.

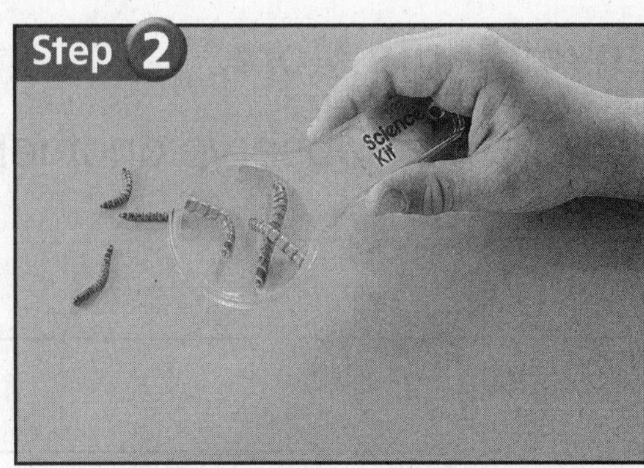

Name ______________________________ Date ____________

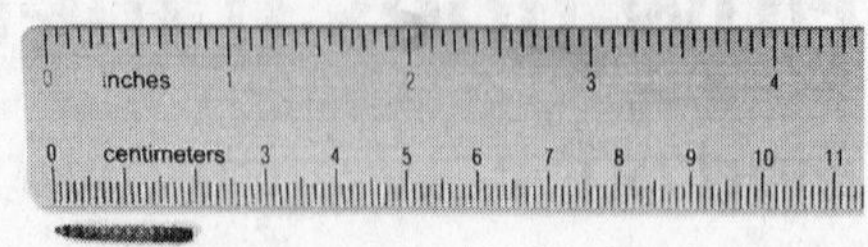

3 **Record Data.** Measure your mealworm every two days. Remember to be gentle with the mealworm. Write about how it changed.

____

____

4 **Predict.** How long do you think your mealworm will grow? How do you think it will change?

____

____

Investigate More

Compare. Observe another mealworm. How are they alike and different?

____

____

____

Name ______________________ Date ____________

Explore

How does the color of an animal keep it safe?

You need

- scissors
- 2 pieces of patterned paper
- stopwatch
- plain paper

What to Do

1. Cut one piece of patterned paper into eight shapes.
2. Put the eight shapes on the other sheet of patterned paper.
3. Time your partner while he or she picks up the shapes.
4. Now put the shapes on plain paper and time your partner again.

Name ______________________________ Date ____________

5 Which was easier to find? Which was faster? Why?

____

____

____

Explore More

6 **Infer.** How would the activity be different if the shapes were placed on solid colored paper?

____

____

____

Name ______________________ Date __________

How can an animal use its environment to stay safe?

You need
- drawing paper
- crayons

In this activity, you will see how animals use their environments to protect themselves.

What to Do

1. On a separate piece of paper, draw a picture with a colorful background. Hide two shapes in your picture. One shape should be the same color as the background, and the other shape should be a different color.

2. Exchange your picture with a partner. Try to find the two shapes in your partner's picture.

3. **Compare.** Which shape was easier to find?

____

____

What Did You Find Out?

4. How can the color of an animal keep it safe?

____

____

____

Quick Lab

Name ______________________ Date __________

Animal Eyes

You need

- paper rolls

1. Choose a partner. One partner should stand in front of the other, facing forward.

2. The student in front should hold up two paper towel rolls to his or her eyes. What do you see?

____

3. The student in back should slowly lift his or her arms around the other student. The partner with the tubes should tell when he or she can see the other partner's hands.

4. Switch places and repeat the activity.

5. **Compare.** Put down the tubes and repeat the activity. Which way made it easier to see your partner's hands?

____

6. **Infer.** Why do you think fish have eyes on the sides of their head?

____

____

Name ______________________ Date __________

Explore

Where do animals live?

What to Do

You need
- paper
- crayons

1. **Observe.** Look at the footprints below. What animal do you think made them?

____

2. **Infer.** How does the shape of its feet help this animal? Share your idea with a partner.

____

____

3. Draw a picture of the animal and the place where it lives.

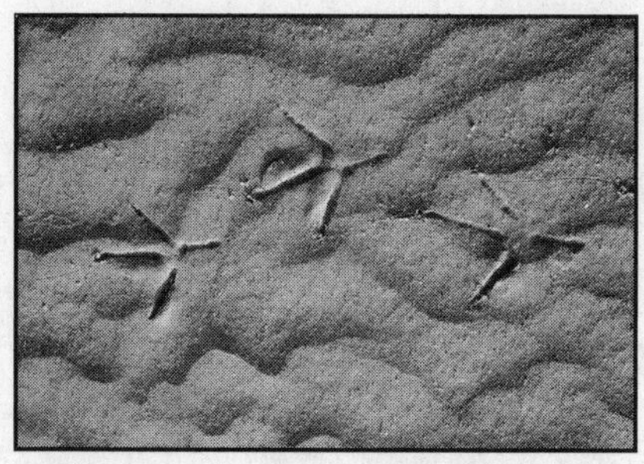

 Name ______________________ Date ___________

Explore More

4 **Communicate.** What other animals could live near this animal? What do they need to live? How do they get food and water? Make a chart.

Name ______________________ Date ____________

How do animals get their needs where they live?

You need

- animal picture

In this activity, you will observe and compare the ways in which animals get what they need in different habitats.

What to Do

1. Observe the animal picture your teacher gives you. Where does it live? What does it eat? What kind of home does it live in?

____

____

____

2. How does the animal take care of its needs in the habitat where it lives?

____

____

3. **Compare.** Show your animal to a partner and compare your ideas. Do animals that live in different habitats take care of their needs in the same way?

____

Quick Lab

Name ______________________ Date __________

Describe a habitat

You need

- nature magazines
- markers

What to Do

1. Look through a nature magazine to find a habitat that you would like to write about.

2. Use markers to draw the living things that you think would live in the habitat you chose.

3. Discuss your habitat with a partner. Do your habitats share any of the same living things?

____

____

Name ______________________ Date __________

Focus on Skills

Put Things in Order

When you put things in order, you tell what happens first, next, and last.

Learn It

Think about how a plant grows. Then look at the pictures and put them in order. You can use a chart to help you tell the order.

The plant gets bigger.

A seedling grows.

I plant a seed.

Focus on Skills

Name ______________________ Date __________

Try It

Look at the pictures below.

beaver dam and pond

beaver cutting trees

stream in the woods

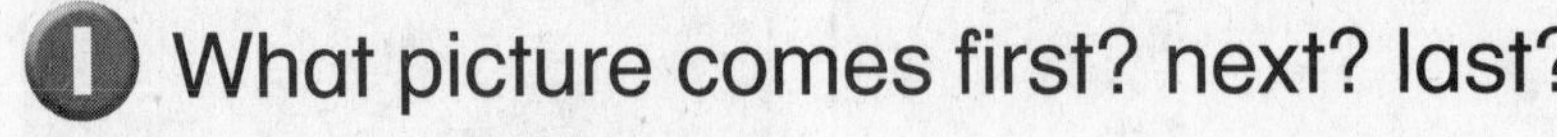

1. What picture comes first? next? last?

____

____

____

2. **Write About It.** Write about what happens to the stream and woods when beavers build a dam.

____

____

____

____

Name ______________________ Date ____________

What do animals eat?

What to Do

1. The Sun makes plants grow. Which animals eat plants? Which animals eat those animals?

____

____

2. Draw the Sun on the yellow strip. Draw some grass and trees on the green strip. Then draw a bird on the red strip and a grasshopper on the brown strip.

3. **Put Things in Order.** Make a chain of strips. Glue them in their order as food.

4. **Communicate.** Describe the order of your chain with a partner.

Name ______________________ Date ____________

Explore More

5 Repeat the activity with three other animals. Communicate how you put the animals in order and draw them below.

____

____

Name ______________________ Date __________

Alternative Explore

What belongs in most food chains?

In this activity, you will work with a partner to explore the differences between food chains.

What to Do

1. Choose a food chain and describe it below.

____

____

2. Compare food chains with your partner. How are they alike? How are they different?

____

____

____

3. What do most food chains have in common?

____

____

 Name ______________________ Date __________

Food chain puppets

You need

- paper plates
- crayons
- glue
- construction paper
- craft sticks

What to Do

1. Work with your group to create a food chain. Which plants and animals will you use in your food chain?

   ____

   ____

2. Use craft materials to make puppets of the plants and animals in your food chain. Draw the plants and animals on the construction paper, then glue them to the paper plates. Glue a craft stick to the plate to complete your puppet.

3. Work with your partners to act out your food chain. The class can guess the plants and animals in your food chain.

Name ______________________ Date __________

Explore

What happens when habitats change?

You need

- large pieces of paper
- crayons
- small toys and blocks

What to Do

1. On a large sheet of paper, draw a large meadow, woods, and river.
2. Place the animals where they would live.

Step 2

3. Use blocks as houses and buildings. Build a town with houses and stores.
4. **Observe.** What happens to the meadow, woods, and animals that live there?

____

____

____

 Name ____________________ Date __________

5 **Infer.** How does building a town affect the animals, meadow, woods, farms, river, and people?

____

____

____

Explore More

6 **Predict.** What will happen if a highway is built?

____

____

____

Name ______________________ Date ____________

What can change habitats?

In this activity, you will draw a habitat and make it change.

You need

- drawing paper
- crayons
- markers

What to Do

1. On a separate piece of paper, draw a picture of land, trees, and small animals that live in a habitat.
2. Draw a town or city over your land with a dark marker. Include roads, houses, and stores.

What Did You Find Out?

3. **Infer.** How did the buildings and roads change the environment?

____

____

____

Quick Lab

Name ______________________ Date ____________

Habitats change

You need
• crayons

What to Do

1. Think of all of the different habitats that you know, and choose one. Which habitat did you choose?

____

2. Describe one way that your habitat can change.

____

____

____

3. Draw a three-panel comic strip showing how your habitat can change. The three panels should show what your habitat looks like before, during, and after it changes.

How do clues help scientists put fossils together?

Find out how scientists put fossils together.

You need

- clay
- leaves
- plastic knife
- hand lens

What to Do

1. Work in a small group. Roll out some clay and press a leaf into it. Peel it off carefully.
2. Cut your leaf print into two pieces. You do not have to use straight lines.
3. Trade your leaf prints with another group.
4. **Infer.** Use clues in the prints to match them and put them together.

Name ______________________ Date ____________

Investigate More

Communicate. How would you put together a dinosaur fossil? How did this activity help you learn how scientists work?

____

____

____

____

Name ______________________ Date __________

Explore

What is a forest like?

What to Do

1. **Make a model** of a forest. Place soil, a plant, and rocks in a bottle.
2. Water the soil. Add a pill bug. Cover the bottle with plastic wrap. Poke holes in it. Place near a window.
3. **Observe** your model. Record on a chart how it changes.

You need

- soil
- plant 
- plastic wrap 
- rocks 
- plastic spoon 
- pill bug

My Model Forest	
Day	**Observations**
1	
2	
3	
4	
5	

Explore

Name ____________________ Date __________

Explore More

4 **Make a model** of the forest in winter. Draw a picture to show how it would change.

Name ______________________ Date __________

Alternative Explore

How do animals live in a forest?

You need
• crayons
• drawing paper

In this activity, you will draw and write about how animals live in forests.

What to Do

1. Draw a picture of a forest.

2. What are three animals that live in the forest?

____

What Did You Find Out?

3. How do the animals in the forest get the things they need to live?

____

____

Quick Lab

Name ______________________ Date __________

Earthworms

You need

- tray
- hand lens
- worm

What to Do

1. Carefully place a worm onto a tray. Use your hand lens to observe the worm's behavior.

2. How does the worm move?

____

____

3. Where do worms live? How can they be an important part of their habitat?

____

____

4. Draw animals that live on the forest floor.

Name ______________________ Date __________

Explore

How does the shape of a leaf help a plant?

You need

- paper towels
- scissors
- tape
- water
- plastic wrap

What to Do

1. Cut two leaf shapes from paper towels.
2. Roll up one leaf shape and tape it closed.

3. Place both leaf shapes on plastic wrap and wet them.
4. **Observe.** Check both leaves every 15 minutes. Which leaf shape stayed wet longer?

____

____

Name ______________________ Date ____________

Explore More

5 **Draw a Conclusion.** Which kind of leaf might you find in a dry place?

____

____

____

Name ______________________ Date __________

Alternative Explore

Where do plants live?

In this activity, you will classify plants by where they live.

You need

- pictures of plants

What to Do

1. **Observe** the photos of plants. Sort the photos into two groups. One group should be desert plants and the other should be rain-forest plants.

2. Compare and contrast the plants that live in the desert with the plants that live in the rain forest. Complete the Venn diagram.

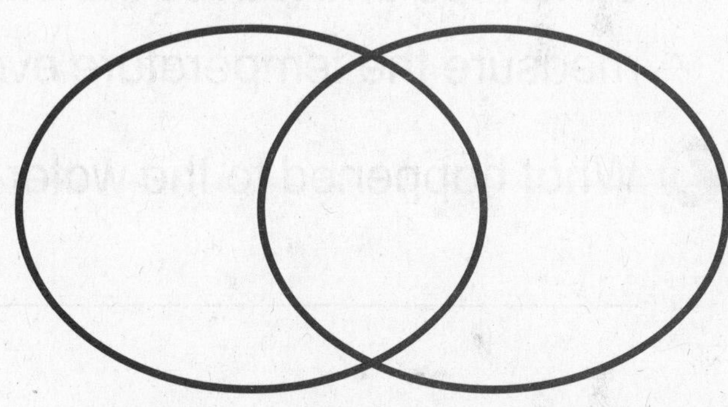

What Did You Find Out?

3. How are the plants that live in the desert different from plants that live in the rain forest?

____

____

____

Quick Lab

Name ____________________ Date __________

Surviving in the Arctic

You need
- hot water
- thermos
- cup
- thermometers

What to Do

1. Observe as your teacher measures the temperature of hot water. What is the temperature of the water?

____

2. After your teacher pours half of the hot water into a thermos and leaves the other half in the cup, measure the temperature every five minutes.

3. What happened to the water in each container?

____

____

____

4. Infer. How does insulation like fur or blubber help animals stay warm?

____

____

____

Name ______________________ Date __________

Infer

You **infer** when you use what you know to figure something out.

Learn It

Lianna and Gene watched each other walk, then run. They made a chart of each other's footprints. Lianna and Gene used what they knew to figure out what the footprints can tell about their steps.

Focus on Skills

Name ______________________________ Date ____________

Try It

What can you figure out about animals from their footprints?

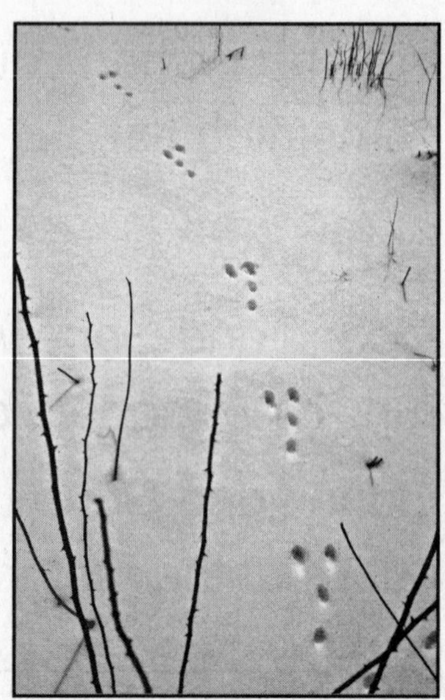

1. Look at the photos of the animal prints above. Tell what you can learn about an animal from its footprints.

____

____

____

2. Share what you infer with a classmate.

3. **Write About It.** Describe how you can use what you know to figure things out.

____

____

Name ______________________ Date ____________

Explore

What lives in a saltwater habitat?

You need

- 2 containers
- salt 

- measuring spoons
- measuring cup
- brine shrimp eggs 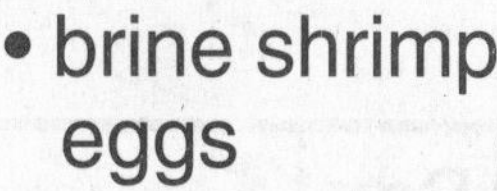
- hand lens  

What to Do

1. Fill each container with two cups of water. Add two teaspoons of salt to one container and stir.

2. Add $\frac{1}{4}$ teaspoon of brine shrimp eggs to each container.

3. **Observe** with a hand lens what happens every day for a week. Record your results in a table.

 Name ______________________ Date __________

Explore More

4 **Record Data.** What happens if you place the shrimp in a dark place for a week?

____

____

____

Day	Observations
1	
2	
3	
4	
5	

Name ______________________ Date __________

Alternative Explore

What lives in a saltwater habitat?

You need

- pictures of animals in water habitats
- paper
- crayons

In this activity, you will classify animals by the kind of water habitat in which they live.

What to Do

1. **Observe** the photos of animals, sort the photos into a group of saltwater animals and a group of freshwater animals.

2. **Record data.** How are the animals that live in salt water like those that live in fresh water? How are they different? Complete the Venn diagram.

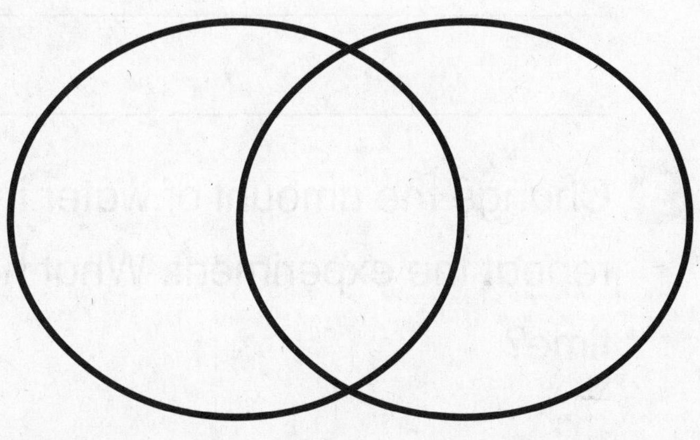

Salt Water Both Fresh Water

What Did You Find Out?

3. Use the information in your Venn diagram to describe how saltwater and freshwater animals are alike.

____

____

Quick Lab Name ______________________ Date ____________

Model a Swim Bladder

You need

- tub of water
- plastic bottle with cap

What to Do

1. Fill a clear tub with water.

2. Fill the bottle halfway with water and screw on the cap. Put the half-filled bottle in the tub of water and observe what happens.

____

____

____

3. Change the amount of water in the bottle and repeat the experiment. What happens this time?

____

____

____

Name ______________________________ Date ____________

How are parts of Earth's surface alike and different?

What to Do

1 Observe. How are the pictures alike? How are they different? Talk about the pictures with your partner.

____

____

____

2 Classify. Sort the pictures into two groups. Describe how you sorted them.

____

____

____

 Name ______________________ Date __________

3 **Classify.** This time sort the pictures into three new groups.

Explore More

4 **Predict.** How might the land change during a year?

____

____

____

Name ______________________ Date __________

How is land different?

You need

- drawing paper
- crayons
- pencil
- magazines

In this activity, you will observe and draw different kinds of land.

What to Do

1. **Observe.** Look through the magazines for pictures of different kinds of land.
2. **Compare.** Talk with your classmates about how the pictures of land are alike and different.
3. Draw a picture showing a type of land. Show what animals, plants, or trees might be on it and how it might be used.

What Did You Find Out?

4. How is land different in different places?

____

____

____

 Name ____________________ Date __________

Make a Model of Earth

You need

- red, orange, and yellow clay
- plastic knives
- crayons

What to Do

1. Roll a small piece of red clay into a ball.
2. Make a flat, thin layer of orange clay and wrap it around the red ball.
3. Using yellow clay, make another flat, thin layer. Wrap it around the orange clay.
4. Use your plastic knife to cut the ball of clay in half. Observe the different layers of clay and draw what you see. How is the ball of clay like Earth?

____

____

____

Name ______________________ Date __________

Make a Model

You **make a model** when you show how something looks or works. Models can help you understand how some things look or work in the real world.

Learn It

A second grade class made this model of mountains and a valley. How do you think this model was made? How does it help you learn about the land?

____

____

____

Name ______________________________ Date ____________

Try It

Make a model to learn more about mountains and islands.

1. Use clay to make a mountain in the bottom of a plastic tub.
2. Slowly pour water into the tub around the mountain. Observe what happens. What type of land does your mountain look like now?

   ____

   ____

3. What does your model tell you about some mountains?

   ____

   ____

Name ______________________ Date ____________

How do people use water?

What to Do

How People Use Water	
drinking water washing dishes swimming	III

1 Observe. How many ways do you use water during the day? Make a list.

____

____

____

2 Communicate. Discuss your list with classmates and add other ways that people use water.

____

____

____

3 Classify. Sort the different ways people use water. How can you classify the different ways?

____

____

____

 Name ____________________ Date __________

Explore More

4 **Investigate.** How many times do you use water in one day? Make a tally chart. Use your tally to make a bar graph.

Name ____________________ Date ____________

Alternative Explore

How is water used?

You need

- pencil
- magazines
- scissors
- glue
- poster board

In this activity, you will explore different ways that water is used.

What to Do

1. Find and cut out pictures showing different ways in which water is used.
2. Glue the pictures onto your poster board and write a label telling how each picture shows water being used.
3. **Communicate**. Share and explain your pictures with a classmate.

What Did You Find Out?

4. How is water used?

____

____

____

5. Why is water important?

____

____

____

Quick Lab

Name ______________________ Date __________

Water and Land

You need

- world map
- crayons

What to Do

1. Observe the world map that your teacher gives you. How is it different from other maps?

____

____

____

2. Predict how much of the Earth is covered by water.

____

____

3. Use your crayons to color in the map. Color water areas blue and land areas brown.

4. Count the number of squares that are land areas and water areas. Was your prediction correct? How close were you?

____

____

Name ______________________ Date __________

How can you change rocks?

What to Do

1. **Observe.** Use a hand lens to look at three or four rocks. Draw what you see.

2. **Predict.** What will happen if you shake the rocks in a jar with water?

____

3. Put the rocks in a jar filled half way with water. Shake the jar for two minutes. Remove the rocks. Observe them again.

Explore

Name ______________________ Date __________

4 **Draw conclusions.** Why do you think the rocks changed or did not change?

____

____

____

____

Explore More

5 **Experiment.** How else can you change the rocks?

____

____

____

____

____

Name ______________________ Date ____________

Alternative Explore

What can change rocks?

In this activity, you will observe how a rock can change.

You need

- pencil
- towel or cloth
- hammer
- various rocks
- sandstone

What to Do

1. **Observe** a rock carefully. Is it rough or smooth?

   ____

2. Watch while your teacher hits the sandstone with a hammer.

3. **Observe.** Look at the sandstone carefully after it has been hit with the hammer.

What Did You Find Out?

4. How did the sandstone change?

   ____

   ____

5. What forces in nature might change a rock like the hammer did?

   ____

   ____

   ____

Quick Lab

Name ______________________ Date __________

Freezing Water

You need
• plastic bottle
• water
• drawing paper

What to Do

1. Fill a small plastic bottle with water and screw on the cap. Draw a picture of the bottle of water and label it "before."

2. **Predict.** What do you think will happen if you leave the bottle in a freezer overnight?

____

____

____

3. Place the bottle in a freezer and leave it there overnight.

4. **Communicate.** What happened to your bottle? Draw a picture of the bottle and label it "after." Was your prediction correct? Share your predictions and results with classmates.

____

____

____

Name ______________________ Date ____________

How can we sort rocks?

You need

- rocks
- hand lens

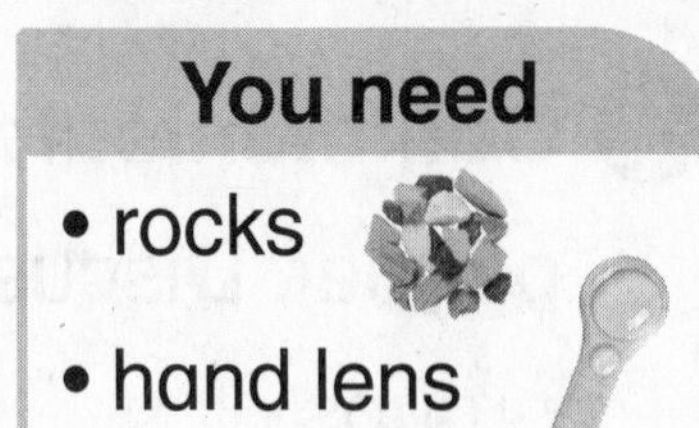

What to Do

1. **Observe.** Look at your rocks with a hand lens. Describe what you see. How are they alike? How are they different?

____

____

____

2. **Classify.** Put your rocks into groups. Write your groups on a chart on a separate piece of paper. Record how many rocks are in each group.

 Name ______________________ Date __________

3 **Communicate.** Share your chart with a partner. Discuss how you put the rocks into groups.

Explore More

4 What other ways can you classify rocks?

___

___

___

Name ______________________ Date ____________

Alternative Explore

Are all rocks alike?

You need

- rocks

In this activity, you and a partner will compare rocks.

What to Do

1. You and a partner should each select a rock. Observe their color, shape, and feel.

2. **Compare.** How is your rock similar to your partner's rock? How is it different?

____

____

What Did You Find Out?

3. How are rocks different from each other?

____

____

4. Why do you think that so many different rocks exist?

____

____

Quick Lab

Name ______________________ Date __________

Observing Minerals

You need
- assorted rocks
- hand lens

What to Do

1. Work with a partner. Choose a rock and observe it with a hand lens.

2. Describe your rock. How many minerals do you see? What colors are the minerals?

____

____

____

3. Work with another pair and compare your rocks. Show how your rocks are alike and different using a Venn Diagram.

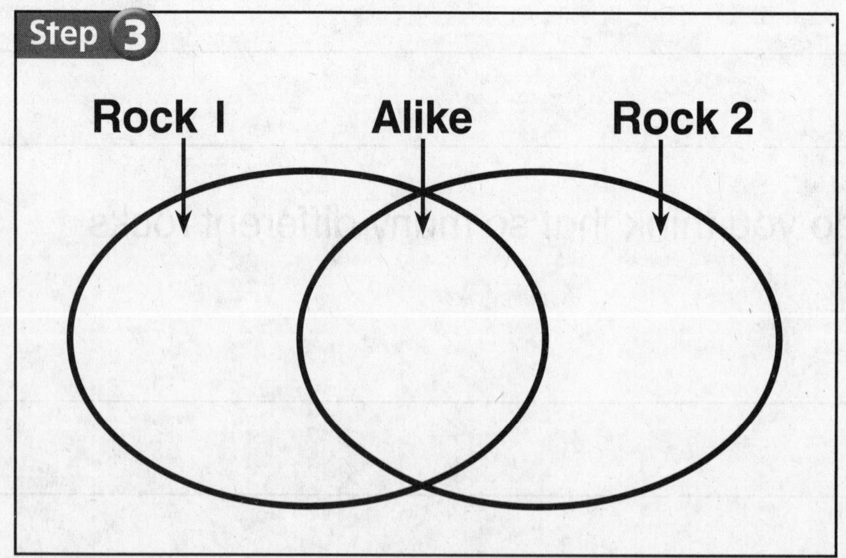

Name ______________________ Date __________

Compare

When you **compare**, you look for ways that things are alike and different.

Learn It

Cats meow and have four legs. Dogs bark and have four legs. You can record how cats and dogs are alike and different in a Venn diagram. Write how the animals are alike in the space where the two circles meet.

Focus on Skills

Name ______________________ Date __________

Try It

feldspar

quartz

1 How are feldspar and quartz alike? How are they different?

____

____

2 Make a Venn diagram to compare feldspar and quartz.

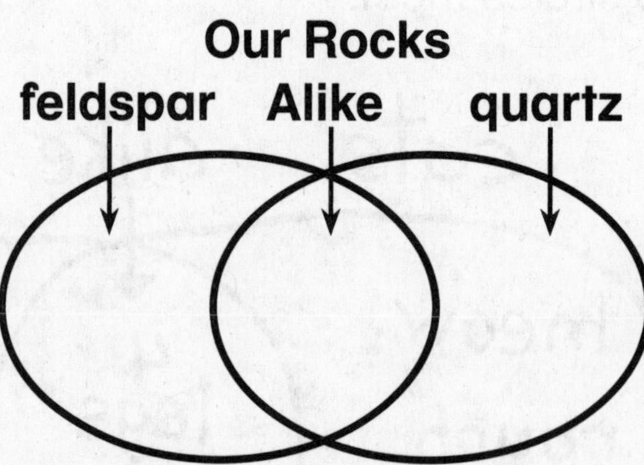

3 **Write About It.** Find two other rocks and compare them. Use a Venn diagram.

Name ______________________ Date __________

Explore

What is in soil?

What to Do

1. Put some soil in a strainer. Gently shake it over a plate.

2. **Observe.** Look at the soil on the plate. Use a hand lens. Draw what you see.

Name ______________________ Date __________

3 Pour the soil left in the strainer onto another plate. Observe the soil. Draw what you see.

Explore More

4 **Communicate.** Use some new soil. Repeat this activity. Write about how the two soils are alike and different.

Name ______________________ Date __________

How are soils alike and different?

You need

- crayons
- pencil
- pictures of soil

In this activity, you will observe and compare different kinds of soil.

What to Do

1. **Observe.** Look carefully at each picture of soil.

2. **Compare.** Talk with a partner about how the pictures are alike and different. Compare the color of the soils, what you see in the soils, and how the soils might feel.

3. **Communicate.** How can the different soils be used? Choose a type of soil. On a separate piece of paper, draw a picture of it being used.

What Did You Find Out?

4. How are soils different?

____

____

____

Quick Lab

Name ______________________ Date __________

Make a Compost Pile

You need

- plastic container
- potting soil
- lunch scraps

What to Do

1. Fill a plastic container halfway with soil. Add your lunch scraps to the soil.
2. Observe the mixed materials in the container. What does the mixture look like? Predict what you think will happen to the materials over time.

____

____

____

3. Each day, add more food scraps and some water to the compost pile. Mix the contents of the container.
4. How has the soil changed after a week?

____

____

____

Name ______________________ Date __________

Be a Scientist

Which soil holds more water?

Find out how different soils hold different amounts of water.

You need

- 2 cups 
- sandy soil
- clay-rich soil
- 2 measuring cups
- clock 
- pencil

What to Do

1. △ **Be Careful!** Use a pencil to poke three small holes in the bottom of each cup. Label the cups A and B.
2. **Measure.** Fill cup A with 1 cup of sandy soil.
3. **Measure.** Fill cup B with 1 cup of clay-rich soil.

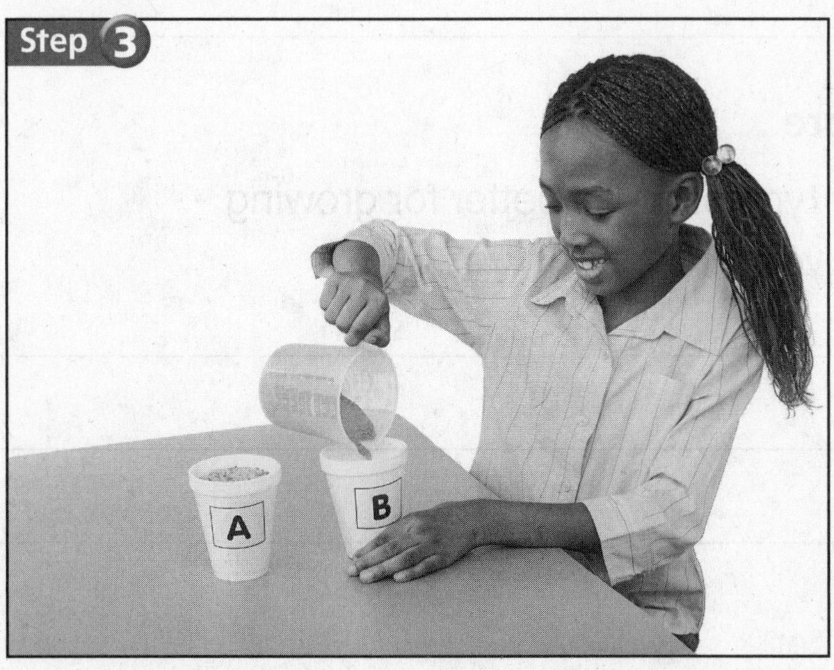

Name ______________________ Date __________

4 **Predict.** Which cup will drip more water from the bottom of the cup? Why do you think so?

____

____

____

5 Hold each cup of soil over a measuring cup. Have a partner pour 1 cup of water into each cup of soil.

6 **Measure.** After 5 minutes, measure how much water dripped into each cup.

____

Investigate More

Predict. Which type of soil is better for growing plants? Why do you think so? Try it.

____

____

____

Name ______________________ Date __________

How do we use Earth's resources every day?

What to Do

1. Make a chart on a separate piece of paper about how you use water, air, plants, animals, and rocks.

2. **Communicate.** Write down your ideas on the chart.

3. Work with a partner. Think of other things you use from Earth. Write down your ideas.

____

____

____

 Name ______________________ Date __________

4 **Draw Conclusions.** How are the things that come from Earth important to us?

____

____

____

Explore More

5 **Infer.** What if there were no more water or rocks on Earth? How would your life change? Write your ideas.

____

____

____

____

____

Name ______________________ Date __________

Alternative Explore

What are things made from?

In this activity, you will find out how we use our natural resources.

You need

- magazines
- paper
- scissors
- pencil

What to Do

1. Read a magazine to find a picture that shows a lot of different objects.
2. **Observe.** Look carefully at your picture. What objects do you see? List the objects in your picture and tell what each one is made from.

____

____

3. **Communicate.** Share your list with your group. What are the objects on the other children's lists made from?

What Did You Find Out?

4. What did you find out about how our natural resources are used?

____

____

____

Quick Lab Name ______________________ Date __________

Reduce, Reuse, and Recycle

You need

- three recycling bins

What to Do

1. Set up three recycling bins labeled "Reuse"; "Plastic and Metal"; and "Paper."

2. Predict how much trash your class can reuse and recycle in one week.

____

____

____

3. Begin filling the bins with trash from your classroom. Things such as rags and egg cartons can be placed in the Reuse bin and used again.

4. After one week of filling the bins, compare your prediction with the actual amount of trash in the bins.

____

____

____

____

Name ______________________ Date ____________

Explore

How does the weather change each day?

You need

- thermometer
- construction paper

What to Do

1. Make a chart with the following columns at the top: Date, Temperature, Weather.

2. **Record Data.** Observe the weather each day. Record what you see. Draw any clouds you see.

 Name ______________________ Date ____________

❸ **Compare.** After several days, compare how the weather changed from day to day.

____

____

____

Explore More

❹ Add a column to your chart called Wind. Record how wind changes from day to day.

Name ______________________ Date __________

How much does the temperature change?

What to Do

1. **Predict.** How much of a difference do you think there will be between inside and outside temperatures?

   ____

   ____

2. Record the temperatures on the class chart for two weeks.

What Did You Find Out?

3. **Infer.** Why do you think the outside temperature changed more than the inside temperature?

   ____

   ____

   ____

 Name ______________________ Date __________

How strong does the wind blow?

You need

- craft sticks
- tissue paper
- tape
- drawing paper
- chenille stick

What to Do

1. Fold about 1 inch of the tissue paper over a chenille stick. Tape the paper down. Bend and tape the chenille stick into a circle.
2. Tape the edges of the tissue paper together to make a tube. Tape a craft stick across the end of the tube where the chenille stick is inside the tissue paper.
3. Go outside with your teacher. Hold your wind tool in the air and observe what happens when the wind blows on it. Draw what your tool looks like on a separate piece of paper.
4. Find another location where you think the wind will have a different strength. Hold up your wind tool and draw what you see.
5. Was the wind stronger in one location than the other? Infer why the wind might have different strengths in different places.

____

____

____

Name ______________________ Date __________

Where did the water go?

What to Do

1. Fill both cups halfway with water. Mark the water levels.

2. Cover one cup with plastic wrap. Tape it to the cup. Place both cups in a sunny place.

You need

- 2 cups
- plastic wrap
- water
- tape
- marker

3. **Predict.** How will the levels of water change in each cup over several days?

____

____

Name ______________________ Date __________

4 **Record Data.** Write what you see in each cup every day.

Day 1	
Day 2	
Day 3	
Day 4	
Day 5	

5 **Draw Conclusions.** What happened to the water levels after several days? Why?

____

____

Explore More

6 What would happen if you used twice as much water? Try it.

____

____

____

Name ______________________ Date ____________

How does water change?

You need

- writing pencil
- colored pencils
- drawing paper

In this activity, you will work with a partner to create a diagram that shows what happens to puddles after a rain storm.

What to Do

1. **Communicate.** Discuss with your partner what you think happens to the water in puddles after a rain storm.
2. Work together with your partner to create a diagram that shows what happens first, next, and last.

What Did You Find Out?

3. What happens to puddles after a storm?

____

____

____

____

____

Quick Lab

Name ______________________ Date ____________

Observe the water cycle

You need

- clear plastic cups
- hot tap water
- plastic wrap
- rubber bands

What to Do

1. Watch as your teacher pours hot tap water into plastic cups.
2. Cover the cups with plastic wrap and secure it with rubber bands.
3. Leave the cups in a warm place. Later in the day, observe the cups and draw a picture of what you see.
4. Discuss the cups with your classmates. Do you see evidence of condensation and evaporation?

Name ______________________ Date ___________

Predict

When you **predict**, you use what you know to tell what you think will happen.

Learn It

Kendra needs to decide which footwear she should wear outside. What do you think she will choose?

What I Know	What I Predict
I know it's raining outside.	I predict that Kendra will wear her rain boots.

Name ______________________ Date ____________

Try It

1. Look through this window and at the thermometer. What type of weather do you think is coming?

____

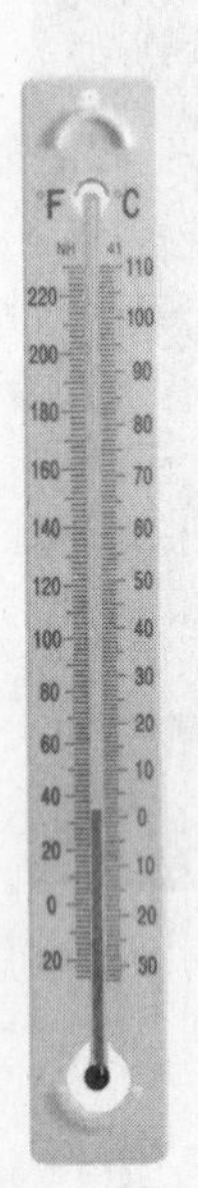

2. What information did you use to help you predict?

____

____

____

3. **Write About It.** What do you need to wear to keep warm on a cold day? Write a story about it on a separate piece of paper.

Name ______________________ Date __________

How can clouds help predict the weather?

What to Do

1. **Observe.** Look carefully at the sky every day this week.

2. **Record Data.** In your notebook, draw the kinds of clouds you see each day for a week. Write the date next to each picture. Then predict what the weather will be like tomorrow.

3. The next day, record what the weather is like. Draw the clouds and the date. Was your prediction from the day before correct?

____

 Name ______________________ Date __________

4 **Draw Conclusions.** How can clouds help predict the weather?

___

___

Explore More

5 **Predict** Write a weather report for the next week. Why is tomorrow's weather the easiest day to predict?

___

___

___

___

___

Name ______________________ Date __________

How accurate are weather predictions?

In this activity, you will compare weather predictions and the actual weather.

You need

- crayons
- drawing paper

What to Do

1. **Record data.** On a separate piece of paper, draw a picture that shows today's weather. Include details that will illustrate the temperature and the amount of wind, clouds, or precipitation that you see.
2. Look at today's weather forecast.

What Did You Find Out?

3. How accurate was today's weather forecast?

____

____

4. **Infer.** What weather conditions do you think the forecasters use to help predict the weather?

____

____

____

Quick Lab

Name ______________________ Date __________

What happens when air moves quickly?

You need

- small brown paper bag

What to Do

1. Watch your teacher blow air into a paper bag and hold it tightly shut. Do the same with your paper bag.

2. Hold the blown-up bag in front of you with one hand, and quickly strike it with the other hand. What happens?

____

____

3. Discuss with your classmates what happens to the air inside the bag when you strike it with your hand.

Name _________________________ Date ____________

Explore

Why can't we see the Sun at night?

You need

- flashlight

What to Do

1. Stand 12 steps away facing a partner.

2. Point a flashlight at your partner. The flashlight is the Sun. Your partner is Earth.

3. **Predict.** Let your partner turn around slowly in front of the flashlight. Will he or she always be able to see the light? Try it.

____

____

____

 Name ______________________ Date __________

4 **Infer.** How does this model show why we can not see the Sun at night?

____

____

____

Explore More

5 **Make a Model.** What pattern is made when your partner turns around in front of the flashlight three times? Try it.

____

____

____

Name ______________________ Date __________

Where is it day and night on Earth?

You need
- globe
- pin
- flashlight

In this activity, you will identify where it is day and night on a globe.

What to Do

1. **Observe.** Shine a flashlight on a globe. Where is it day? Where is it night?

____

____

____

2. **Make a model.** Find your hometown on the globe and mark it with a pin. Use the flashlight to show your town during the day. Then rotate the globe to show your town at night.

What Did You Find Out?

3. What causes the change from day to night and back to day again?

____

____

____

Quick Lab

Name ______________________ Date __________

Make a flip book

You need

- index cards
- markers
- stapler

1. On each index card, draw a horizon line about an inch from the bottom of the card. Stack the cards and staple the top two corners.

2. Draw a moon on the left side of the first card. Draw the same moon on each card, moving it slightly to the right every time.

3. Flip through the book. Observe how the Moon seems to move across the sky. What does this movement look like?

____

____

____

Name ______________________ Date ____________

Inquiry Skill: Draw Conclusions

Learn It

When scientists **draw conclusions**, they use what they observe to explain what happens.

Linda looks at this picture.

She sees the lights on and the dark sky. Linda has seen some of the houses before. She draws the conclusion that this picture was taken at night in her town.

Try It

Observe the lengths of shadows. Then draw conclusions about the time of day.

1. Push a stick straight into a pot of dirt. Place the pot in a sunny spot.

Focus on Skills

Name ______________________ Date __________

2 Look at the stick at different times of day. Sit in the same spot each time. Draw the Sun, stick, and shadow. Write the time of day on each drawing.

3 Compare. Talk to your partner about how the shadows changed. When was the shadow longest?

____

____

____

4 Draw Conclusions. What does the time of day have to do with the length of shadows?

____

____

____

____

Name ______________________ Date __________

Explore

What clothes do people wear in each season?

You need

- paper
- markers
- magazines
- scissors
- glue stick

What to Do

1. Write the name of a different season in each corner of your paper.
2. Cut out pictures of different kinds of clothes from magazines.

3. **Classify.** Glue the pictures near the seasons where they belong.
4. **Draw Conclusions.** What do people wear in different seasons?

____

____

____

Name ______________________ Date __________

Explore More

5 **Classify.** Sort your clothes at home by season. Explain how you grouped your clothes.

____

____

____

Name ____________________ Date __________

Alternative Explore

What clothing do you wear during a season?

You need
- pencil

In this activity, you will list clothing you wear during each season.

What to Do

1. What kind of clothes do you wear during each season? Discuss it with a partner and record your ideas in the boxes in the chart.

Spring	Summer	Fall	Winter

2. **Record data.** Copy your lists from this worksheet onto a large class graph.

What Did You Find Out?

3. **Draw Conclusions.** What does the class graph show you about seasons where you live?

____

____

____

Quick Lab Name ____________________ Date __________

Make a seasons plate

You need

- paper plates
- crayons
- brass fasteners
- scissors

1. Divide a paper plate into four sections by drawing a straight line down the middle, and then another line going across.
2. In each section, draw a picture of an activitiy that you like to do in a season. Make sure you use all four seasons.
3. Take a second paper plate and divide it into four sections. This time, use scissors to cut one of the sections out.
4. Attach the second plate to the top of the first one by sticking a brass fastener through both plates. Turn the top plate to expose an activity and have a partner guess the season.

Name ______________________ Date ____________

How do we see the Moon at night?

You need

- flashlight
- white ball

What to Do

1. Use a white ball as the Moon. Turn out the room lights. Is it easy to see the Moon?

____

____

____

2. **Make a Model.** Shine a flashlight on the Moon. The flashlight is the Sun. Is the Moon easier to see now? Why?

____

____

____

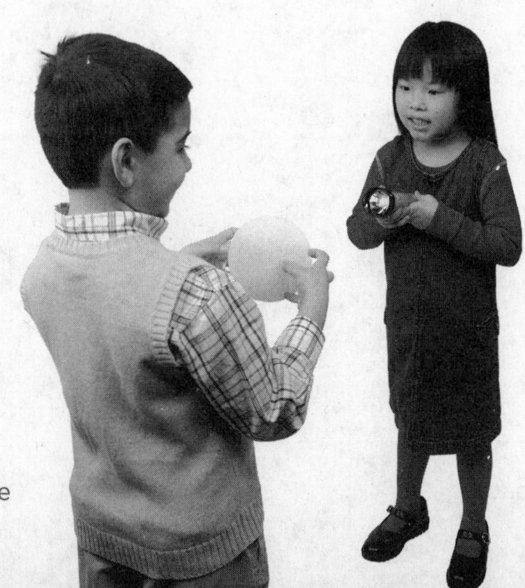

Explore

Name ______________________ Date ____________

3 Draw Conclusions. Where does the Moon's light really come from?

____

____

____

Explore More

4 Investigate. What if the Moon were a different color? How would that affect the brightness of the Moon? Make a model to find out.

____

____

____

Name ________________________ Date __________

What makes the Moon visible from Earth?

You need

- light-colored rock
- medium-colored rock
- dark-colored rock
- black paper
- flashlight

In this activity, you will compare different colored rocks against a dark background to model the Moon in the night sky.

What to Do

1. **Compare.** Place rocks of different shades on a piece of black paper and see which rocks blend in and which stand out.

2. Turn out the lights. Talk with a partner about how easy or hard it is to see the rocks.

3. Stand far away from the rocks and shine a flashlight on them. Which rock is easiest to see?

____

What Did You Find Out?

4. Draw Conclusions. Which rock is most like the Moon? Why?

____

____

Quick Lab

Name ______________________ Date __________

A look at stars

1. **Observe.** With an adult, go outside at night and observe the sky. What date and time are you observing the sky? Draw what you see.

____

2. What shape was the moon? How many bright stars were there?

____

____

3. Share your observations with a classmate.

Name ______________________ Date __________

Be a Scientist

How does the Moon seem to change during one month?

You need

- calendar
- markers

Find out how the Moon seems to change shape each week.

What to Do

1. **Observe.** Look outside tonight. Find the Moon in the night sky.

2. **Record Data.** Draw what the Moon looks like on today's date on the calendar.

3. Repeat steps 1 and 2 each night for a month.

Name ______________________ Date __________

4 When did you see a Full Moon during the month? When did you see a New Moon?

____

____

____

5 **Draw Conclusions.** What do your drawings tell you about the phases of the Moon?

____

____

Investigate More

Predict. How do you think the Moon will look in the sky during the next month? Test your idea. Compare it to the calendar for this month.

____

____

____

Name ______________________ Date __________

How are orbits alike and different?

You need

- poster paper
- crayons
- ruler

What to Do

1. Draw a Sun in the middle of poster paper.
2. **Measure.** Draw an X 6 centimeters to the right of the Sun. Measure another 6 centimeters from that spot. Draw another X.
3. **Make a Model.** Draw a path around the Sun for each X. Each path shows an orbit.

Name ______________________ Date __________

4 **Draw Conclusions.** Which orbit is larger? How do you know?

____

____

____

5 **Make a Model.** Continue drawing Xs until you have 8. Show which orbit is largest.

____

____

____

Name ______________________ Date __________

Alternative Explore

How long does it take to complete an orbit?

You need

- chair
- measuring tape
- masking tape

In this activity, you will compare how long it takes to complete different orbits around the same object.

What to Do

1. **Measure.** Place a chair in the middle of the room. Use masking tape to make four circles of different sizes on the floor around the chair. The circles should be about one yard away from each other.

2. Have your partner stand in place on one of the circles. Starting next to your partner, walk heel to toe along the circle until you reach your partner again. Count your steps. How many steps did it take to orbit the chair?

____

What Did You Find Out?

3. Compare. Record the number of steps it took to orbit each circle.

   Circle 1: __ Circle 2: __ Circle 3: __ Circle 4: __

Quick Lab

Name ______________________ Date ____________

Make a model of the solar system

You need

- paper plates
- string
- safety scissors
- tape
- crayons
- paper

1. Draw a straight line across a paper plate. Along this line, poke ten holes in the paper plate.
2. Cut out a large circle for the sun. Then cut out four medium circles for the four largest planets, and five small circles for the five smallest planets. Color and label each planet, and tape each planet to a piece of string.
3. Put the Sun and the eight planets in order, and hang them from the plate by putting the strings through the holes you poked and taping down the strings.

Name ______________________ Date __________

How can you describe objects?

You need

- crackers

What to Do

1 **Observe.** Look at each cracker. Think about the different ways you can describe the crackers. What words can help you describe each one?

____

____

____

2 **Record Data.** Make a chart on a separate piece of paper like the one shown. Write your observations on your chart.

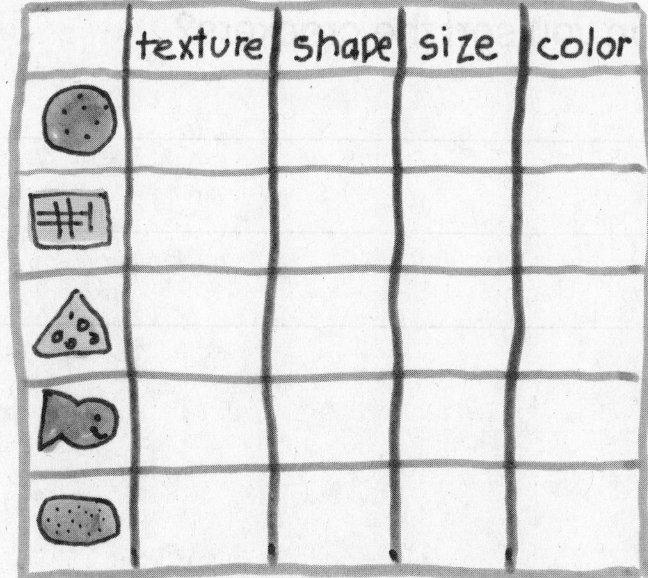

Name ______________________ Date ____________

Explore More

3 **Classify.** Use your chart to help you sort the crackers.

4 How else can you sort the crackers?

____

____

____

Name ______________________ Date __________

Alternative Explore

How are objects alike and different?

You need

- pencil
- paper

In this activity, you will use a Venn diagram to record how two objects are alike and different.

What to Do

1. Look around the classroom and select two different objects that are the same color.
2. On a separate piece of paper, draw a Venn diagram. Label the Venn diagram with the name of each object. In each oval, list some properties of one of the objects. In the middle section, list the properties that both objects have in common.
3. Ask a partner to check your work. If they can think of more ways to describe each object, add their ideas to your diagram.

What Did You Find Out?

4. How did you determine which objects were alike and which ones were different?

____

____

Quick Lab

Name ______________________ Date __________

Size and Shape

You need

- crayons
- classroom objects
- paper

What to Do

1. Observe six objects in the classroom.
2. Take a piece of paper and fold it in thirds. Write Object on the top of the first fold, Shape on the top of the second, and Size on the top of the third.
3. Write the name of the objects in the first column. Draw the shape of each object in the second column and describe each object's size in the third column.
4. Classify the objects by size and compare your findings with fellow students.

Name ______________________ Date __________

Record Data

When you **record data**, you write down what you observe.

Learn It

Joanie talked to each of her classmates about what they had for lunch. She made a tally chart to help her count the kinds of foods they ate. She recorded what was a liquid and what was a solid.

Our Lunch

liquid	IIII
solid	~~IIII~~ III

Then she made a bar graph from her results. A bar graph is a good way to compare data in different groups.

Name ______________________ Date ____________

Try It

Look at this picture. Some things are natural and some are made by people. Make a tally chart to show how many of each thing you see. Then display your data in a bar graph.

1. How many things in the picture were made by people?

____

____

2. What kind of chart can help you record your data?

____

____

3. **Write About It.** How can a bar graph help you compare data?

____

____

____

Name ______________________ Date ____________

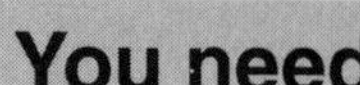

What are the properties of these solids?

You need

- spoons 
- tub of water 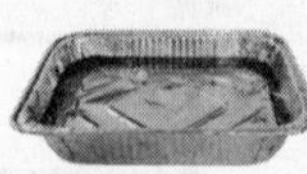

What to Do

1. **Observe.** Look at each spoon. What are the properties of each?

   ____

   ____

   ____

2. **Predict.** Which spoons will float in water? Which will sink? Try it out.

   ____

   ____

Explore Name ______________________ Date ____________

3 **Record Data.** Make a chart to list what you observe.

Type of Spoon	
Sink	**Float**

Explore More

4 **Predict.** How will your list change if you use different objects? How can you find out?

____

____

____

____

Name ______________________ Date __________

Alternative Explore

What happens when solids fall?

In this activity, you will investigate what happens when you drop three different solids.

You need

- pencil
- eraser
- paper

What to Do

1. Drop a pencil onto your desk. What happened?

   ____

2. Drop an eraser onto your desk. What happened?

   ____

3. Drop a piece of paper onto your desk. What happened?

   ____

What Did You Find Out?

4. Why did these solids behave differently when you dropped them?

   ____

   ____

   ____

Quick Lab

Name ____________________ Date __________

Measuring Mass

You need

- balance
- classroom objects

What to Do

1. Look around the classroom and select some objects that you can measure with a balance.

2. **Measure.** Choose two items that you think have the same amount of mass. Measure the mass of each with a balance. Which object had more mass?

   ____

   ____

3. **Compare.** Work with a partner. Keep the original two objects, while your partner tries to find a third object that has equal mass. Put the three objects in order from lightest to heaviest. Switch roles and repeat the experiment.

4. Was it easy to find a third object with the same mass? Why or why not?

   ____

   ____

What happens to water in different shaped containers?

What to Do

1. Put the containers on a tray. Measure one cup of water with the measuring cup. Pour the water into the first container. Mark where the water stops.

2. **Predict.** How high will the same amount of water be in the other containers?

   ____

   ____

   ____

3. Pour one cup of water into the next container. Mark where the water stops. Repeat for each container.

 Name ______________________ Date ____________

4 **Draw Conclusions.** Were your predictions correct? Explain.

____

____

____

Explore More

5 **Infer.** Would the activity change if you used juice instead of water? Why or why not?

____

____

____

Name ______________________ Date __________

How high will the liquid go?

In this activity, you will observe how water level is affected by the shape of a container.

You need

- three different-shaped containers
- measuring cup
- water
- marker

What to Do

1. Observe three different-shaped containers. Predict how high the water level will go when you pour one cup of water into each container. Mark the spot on each container where you think the water will rise to.

2. Measure and pour 1 cup of water into each container. Make another mark to record the actual water level.

What Did You Find Out?

3. Were your predictions correct? How does the shape of a container affect the height of the water level?

____

____

____

Quick Lab

Name ______________________ Date __________

Classifying Matter

You need

- 6 cans with lids
- 2 different solids
- 2 different liquids

What to Do

1. Place a different solid in each of 2 cans, and a different liquid in each of 2 cans. Leave the other 2 cans empty. Label the cans from 1 to 6.
2. Have a partner gently shake each can to infer what type of matter is inside. Sort the cans into solids, liquids, and gases. Complete the chart below by checking the box under the type of matter for that can.

Can	Solid	Liquid	Gas
1			
2			
3			
4			
5			
6			

3. Open the cans and check your results.

Name ______________________ Date __________ **Explore**

How can clay be changed?

You need

- modeling clay 
- balance
- plastic knife

What to Do

1. **Measure.** Find two pieces of clay that are the same mass. Use a balance to show they are equal.

2. Sqeeze and shape one piece of clay into a ball. Describe its properties.

____

____

____

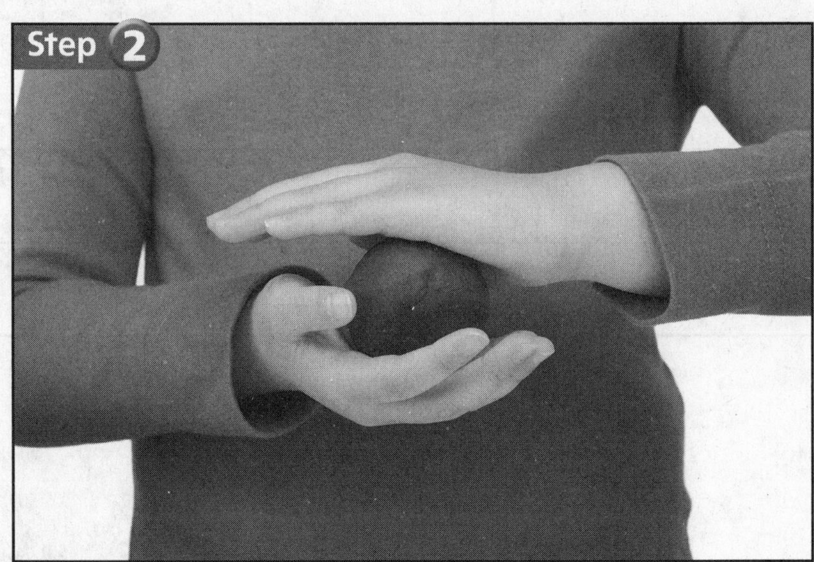

3. **Predict.** Do you think the mass of the clay changed after it was made into a ball? Place it back on the balance to find out.

Explore

Name ______________________ Date ____________

4 △ **Be Careful!** Cut the clay ball into two halves with a plastic knife. Make the two pieces into two figures.

5 **Draw Conclusions.** How did you change the clay?

____

____

____

Explore More

6 **Investigate.** What other ways can you change clay? Will the mass change?

____

____

____

Name ______________________ Date __________

What is the mass of water?

In this activity, you will experiment to find out if the mass of water changes when it is held in different-sized containers.

You need

- 2 containers of different sizes
- 2 measuring cups
- balance

What to Do

1. Put two containers of different sizes on a balance. If they do not have the same mass, add weight to one side until the balance is level.

2. Remove the containers from the balance and fill them both with exactly 1 cup of water.

3. **Predict.** What will happen when you put the containers back on the balance?

____

What Did You Find Out?

4. What effect does the shape of a container have on the mass of water?

____

____

Quick Lab

Name ______________________________ Date ____________

Observe a Chemical Change

You need

- apple
- lemon juice
- plastic wrap
- plates

What to Do

1. Pour lemon juice over 2 apple slices and cover them with plastic wrap. Put them on a plate with two uncovered slices and wait for an hour.

2. After an hour has passed, observe the apple slices. Compare the slices that were covered with the slices that were not.

____

____

____

3. What do you think caused the chemical change to the uncovered apple slices? Do you think you can make the uncovered slices look like the covered slices?

____

____

____

Name ______________________ Date ____________

Communicate

You **communicate** when you draw, write, or share your ideas with others.

Learn It

Joanne changed a ball of clay. She wrote a list to show others how she changed it.

Changing Clay

1. I rolled the clay.
2. I pinched the clay.
3. I squeezed the clay.
4. I poked the clay.

Focus on Skills

Name ______________________ Date __________

Try It

How many ways can you change a piece of paper?

1. Use a chart like Joanne's to communicate how you changed the paper.

2. Share your chart with a classmate.

3. **Write About It.** Tell how your charts are alike and how they are different.

____

____

____

Name ______________________ Date ____________

What happens when you shake cream?

Find out what will happen to cream when you shake it.

You need

- measuring cup
- cream
- jar
- crackers

What to Do

1. **Measure.** Put one quarter cup of cream into the measuring cup.
2. Pour the cream into the jar. Put the lid on tightly.
3. Take turns shaking the jar.

Name ______________________ Date __________

4 **Observe.** What happened to the cream? How did it change? Put it on a cracker.

____

____

____

5 **Draw Conclusions.** How do we use cream? Discuss your answers with a partner.

____

____

____

Investigate More

Observe. What happens if you freeze the cream while shaking it?

____

____

____

Name ______________________ Date __________

How can heat change matter?

You need

- paper plates 
- butter 
- chocolate

What to Do

1. **Predict.** What do you think will happen to butter and chocolate in sunlight?

____

____

2. **Observe.** Place the butter and chocolate on two plates. Draw how they look.

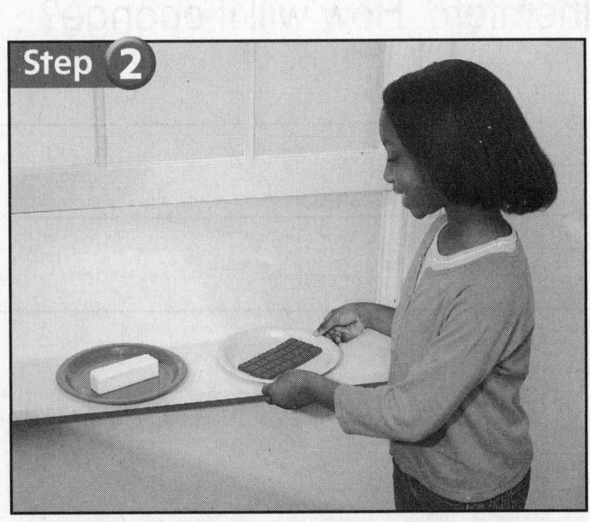

Explore

Name ______________________ Date ____________

❸ **Predict.** How will the Sun's heat change each thing? Find a sunny spot. Leave the plates in the sunlight.

❹ **Communicate.** What happens to each thing after one hour? Draw how they look. Compare your pictures.

Explore More

❺ Now try another item. How will it change?

____

____

____

Name ______________________ Date __________

How does matter change?

You need
- pencil
- colored pencils

In this activity, you will work with a partner to explore what happens to substances in a solid state when heat is added.

What to Do

1. Discuss with your partner the differences between an ice cream cone and a marshmallow.

2. What would happen if you left both on the table?

____

____

3. Predict what would happen if you held a marshmallow over a fire.

____

What Did You Find Out?

4. **Infer.** Why do some solids melt differently?

____

____

Quick Lab

Name ______________________ Date __________

Water Changes States

You need

- magazines
- construction paper
- scissors
- glue
- markers

What to Do

1. Look through magazines and cut out pictures of water in all three states.
2. Glue the pictures of water to construction paper to make a collage. Make sure to have pictures of water as a solid, liquid, and gas.
3. Since water vapor is invisible, how can you illustrate pictures of water as a gas?

____

____

____

____

____

Name ______________________ Date __________

What mixes with water?

What to Do

1 Measure. Add $\frac{1}{4}$ cup salt to one cup of water. What happens?

____

____

____

You need

- measuring cup 
- 2 plastic cups 
- 2 spoons 
- salt 
- sand

2 Measure. Add $\frac{1}{4}$ cup sand to another cup of water. Does the sand change?

____

____

____

 Name ______________________ Date ____________

3 Compare. Stir both mixtures with a spoon. Let them sit. What happens? How are the mixtures different from each other?

____

____

____

Explore More

4 Investigate. Tell how you could take the sand and the water apart. Can the salt be taken out of the water?

____

____

____

____

Name ______________________ Date __________

What difference does temperature make?

You need

- pencil
- 2 jars
- salt

In this activity, you will discover whether salt dissolves faster in hot or cold water.

What to Do

1. Will salt dissolve faster in hot or cold water?

____

____

2. Observe your teacher shake two jars, one with hot salt water, and one with cold salt water.

3. **Compare.** Compare the two jars. What happened? Was your prediction accurate?

____

____

What Did You Find Out?

4. **Infer.** Why do you think the salt dissolved faster in hot water?

____

____

Quick Lab

Name ______________________ Date __________

Evaporation Separation

You need

- salt water
- shallow container

What to Do

1. Pour some salt water into a shallow container. Pour just enough salt water to fully cover the bottom of the container.

2. Predict what will happen to the salt water overnight.

____

____

____

3. Check the container the next day. What happened to the salt water? Was your prediction correct?

____

____

____

Name ______________________ Date __________

Explore

What words help us find things?

What to Do

1. Work with a partner. Pick an object in the classroom. Do not tell your partner what the object is.

2. **Communicate.** Describe where your object is. Give clues to your partner. Ask your partner to find the object.

3. Switch with your partner and try again.

4. **Draw Conclusions.** Which words in your description were most helpful to your partner?

____

____

____

Explore

Name ______________________ Date __________

Explore More

5 **Communicate.** Draw a picture and write directions to find an object in your picture. Then switch with a partner.

Name ______________________ Date __________

Can you get there?

You need

- pencil

In this activity, you will create and follow directions that lead to a mystery place.

What to Do

1. Choose a mystery place for a partner to find. Write three directions to tell how to move to the place. Use position words.

   ____

   ____

   ____

2. Read the clues to your partner. Ask your partner to follow the directions.

3. After your partner finds the mystery place, the activity can be repeated. This time your partner can choose the mystery place.

4. Which words helped you to find the place?

   ____

   ____

Quick Lab Name ________________ Date __________

Speeds of Walking and Hopping

You need

- stopwatch
- tape
- measuring sticks
- open space

What to Do

1. **Predict.** Which is faster: walking or hopping?

   ____

   ____

2. **Measure.** Have a partner help you measure three meters in length on the ground. Use tape to mark the distance.

3. **Record Data.** Draw a two-column chart on a separate piece of paper. Label the columns Walking and Hopping. Ask your partner to first walk and then hop the three meter length. Use the stopwatch to time how long each trip takes. Repeat the activity. This time, your partner will time how fast you can walk and hop. Which was faster?

   ____

   ____

Name ______________________ Date __________

Investigate

When you **investigate**, you make a plan and test it out.

Learn It

Joe and Pat will run in a race. They want to find their speeds. They make a plan.

First, they measure 20 meters. They make a start and a finish line. Next, they measure the time it took them to run the distance.

Look at the chart. Who is faster?

Our Race	
Joe	30 seconds
Pat	28 seconds

Name ______________________ Date __________

Try It

You need

- masking tape
- ruler
- windup toys
- stopwatch

Which toy moves fastest? Make a plan to find out. Then test your plan.

1. Put tape on the floor to make a start line. Measure how far away your finish line will be. Mark it with tape.

2. Use a stopwatch to find out how long it took each toy to go the distance. Record the times.

____

____

____

3. Which toy was fastest?

____

Name ______________________ Date ____________

How do you make things go farther and faster?

You need

- toy car 
- masking tape 
- ruler

What to Do

1. Line up the car at a starting line. Push the car gently over the line.
2. **Measure.** How far did it go?

____

____

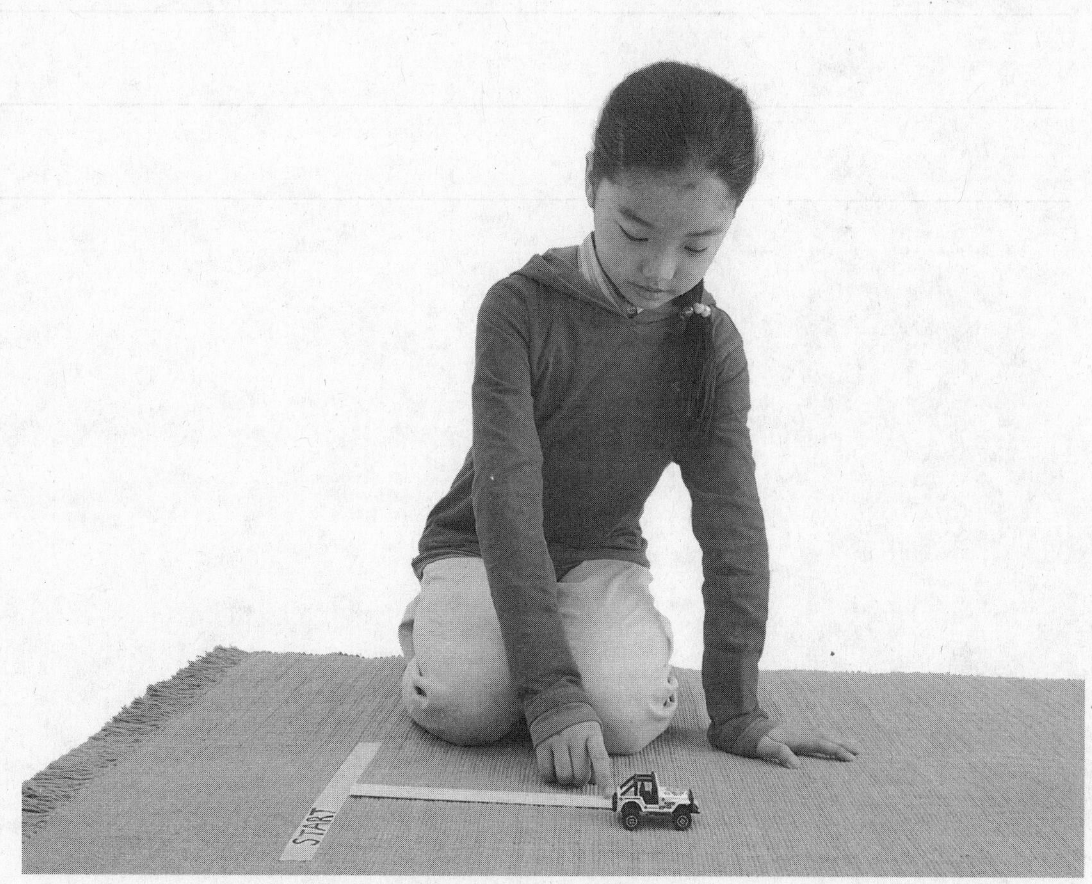

 Name ______________________ Date ___________

3 Do the activity again, but this time push the car harder. Observe what happens.

Explore More

4 **Predict.** What might happen if you pulled the car toward you with your hands? Would it go as far?

____

____

____

____

____

Name ____________________ Date ____________

Alternative Explore

How much force is needed?

In this activity, you will compare forces by pulling loads with a rubber band and measuring the lengths of the stretched band.

You need

- string
- 2 books
- large rubber band
- ruler
- safety goggles

What to Do

1. Tie a piece of string around one book.
2. Tie the string to a rubber band.
3. **Record data.** Pull the rubber band slowly. Record its length at the moment the book begins to move.

   ____

4. Tie two books together, attach the same rubber band and repeat the activity.

   ____

What Did You Find Out?

5. **Infer.** What does the difference in the rubber band length tell you about the force needed to pull the loads?

   ____

   ____

   ____

Quick Lab

Name ____________________ Date __________

Friction and Moving Objects

You need

- stopwatch
- wood blocks or other flat objects with smooth surfaces
- cardboard
- sheets of sandpaper
- books

What to Do

1. With a partner, construct a ramp using books and cardboard.
2. **Record.** Let your partner place a small wood block on the top of the ramp and then let it go. Use a stopwatch to record the time it takes for the block to slide down the ramp.
3. Cover the ramp with sandpaper and repeat the activity.
4. **Compare.** Did the block slide down the ramp faster with or without the sandpaper? How did friction affect the speed?

____

____

____

Name ______________________ Date ____________

Explore

Which side will go up?

What to Do

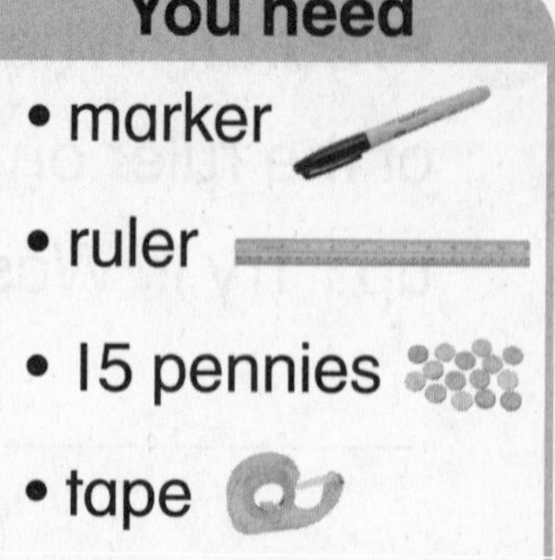

1. Tape a marker to the middle of your desk.

2. Tape 10 pennies to the edge of one end of a ruler. Tape 5 pennies to the edge of the other end.

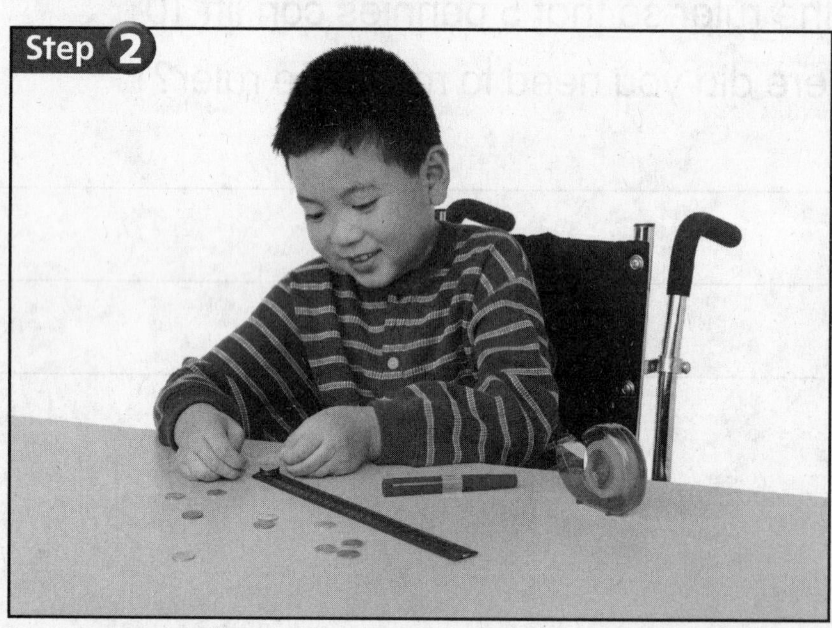

Name ______________________ Date __________

3 **Predict.** What will happen if you put the middle of the ruler on the marker? Which side will lift up? Try it. Was your prediction correct?

____

____

Explore More

4 Try to move the ruler so that 5 pennies can lift 10 pennies. Where did you need to move the ruler?

____

____

____

How can you make a lever?

In this activity, you will lift a hardcover book with a lever made from a marker and a ruler.

You need

- marker
- hardcover book
- tape
- ruler

What to Do

1. Tape a marker to the table. Place the ruler across the marker.
2. Place a book on one end of the ruler, then lift the book by pushing down on the other end of the ruler.
3. Position the ruler at different points to observe where the lever lifts the book highest.

What Did You Find Out?

4. **Observe.** Where was the best place to position the ruler?

____

____

____

Quick Lab

Name ______________________ Date ____________

Making a Pulley

You need

- rolling pin
- yarn
- small pails
- toy blocks

What to Do

1. Along with a partner, place toy blocks in a pail. Tie one end of the yarn to the handle of the pail. Place the pail on the floor and drape the yarn over the center of the rolling pin.

2. Ask your partner to hold on tightly to both ends of the rolling pin. You can now pull down on the yarn to lift up the pail!

3. **Compare.** Try lifting the pail without the pulley. Was it easier to lift the pail with or without the pulley?

____

____

Name ________________________ Date ____________

What can a magnet pick up?

What to Do

1. **Predict.** Put the objects in a bag. Which objects will stick to a magnet?

____

____

____

2. Tie a string to a pencil. Tie a magnet to the end of the string.

You need

- small objects 
- paper bag 
- string 
- pencil
- magnet 

 Name ______________________ Date __________

3 Use the magnet to pull out objects from the bag.

Explore More

4 **Classify.** How are the things that stick to the magnet alike?

____

____

____

____

Name ______________________ Date __________

Alternative Explore

What metals are attracted to magnets?

In this activity, you will separate non-magnetic aluminum from magnetic metals by sorting metal objects for recycling.

You need

- metal objects such as soda cans and aluminum foil
- magnet

What to Do

1. Work with a partner. Gather several pieces of metal and aluminum provided by your teacher.
2. **Investigate.** Test each object to see if it can be pulled by a magnet.
3. **Classify.** Sort the objects that can be pulled by the magnet in one group. Put the ones that can not be pulled by a magnet in another group.

What Did You Find Out?

4. Which objects cannot be pulled by a magnet?

____

____

____

 Name ____________________ Date __________

The North and South Poles of a Magnet

You need

- bar magnets
- sticky notes

1. Observe with your partner how two bar magnets react to each other.

2. Cover the poles on the magnets using the sticky notes.

3. **Experiment.** Try to find the north and south poles of each magnet and then take off the labels to see if you were right. How were you able to identify the north and south poles?

____

____

____

____

____

Name ______________________ Date ____________

Be a Scientist

How can you compare the strength of different magnets?

You need
- paper clips
- magnets

Find out how many paper clips each of the magnets can attract.

What to Do

1. Hang a paper clip from a magnet. Keep adding more clips in a line until no more will stick.

2. **Record Data.** Write how many paper clips can hang from the magnet.

____

____

Name ______________________ Date __________

3. Repeat the steps using different magnets.

4. **Communicate.** Make a bar graph to show the strengths of your magnets. Use the graph below as an example.

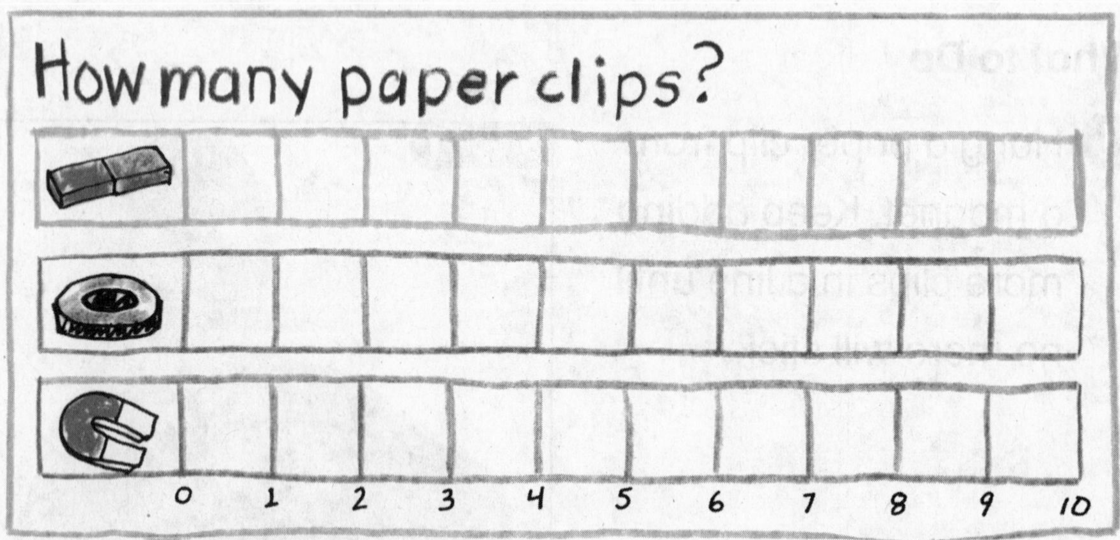

Investigate More

Investigate. How many paper clips can you pick up with two magnets? Find a way to attach two magnets and try it out.

____

____

____

Name ______________________ Date __________

Explore

Where will ice cubes melt more quickly?

What to Do

1 Fill two cups with equal amounts of ice. Place one cup in a sunny place. Place the other cup in a shady place.

You need

- ice cubes 
- 2 cups 
- watch or clock 

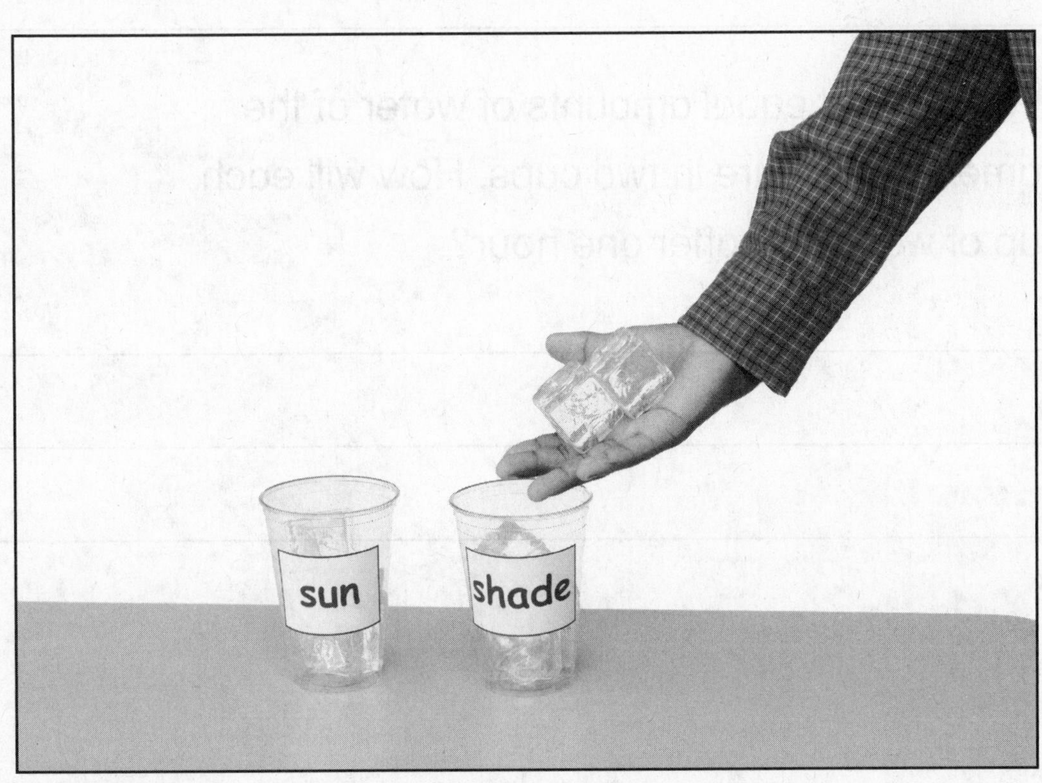

2 **Predict.** Which cup of ice will melt first?

____

____

____

Explore Name ______________________________ Date ____________

3. Record how long it takes for the ice in each cup to melt. Why did one cup of ice melt more quickly?

____

____

____

4. **Predict.** Put equal amounts of water of the same temperature in two cups. How will each cup of water feel after one hour?

____

____

____

Name ______________________ Date __________

How can ice be melted quickly?

You need

- ice
- plate

In this activity, you will discover what happens when you add body heat to ice.

What to Do

1. **Predict.** What do you think will happen if you and your classmates pass around an ice cube?

   ____

2. Using your hands, pass around an ice cube for at least five minutes. Place another ice cube on a plate. Leave it on the plate as you pass around the other ice cube.

What Did You Find Out?

3. Did the ice cube on the plate and the one you passed around melt at the same rate? Why or why not?

   ____

   ____

   ____

   ____

Quick Lab Name ______________________ Date ____________

Test soil, water, and air temperatures

You need

- 3 plastic cups
- thermometer
- paper
- water
- soil

1. Take one plastic cup and put some soil in it. Put water in another cup. Leave the third cup empty.

2. **Predict.** Which cup do you think has the lowest temperature? Which has the highest temperature?

____

____

____

3. **Observe.** Put a thermometer into each cup and read the temperature. Be sure to let the thermometer return to room temperature before putting it in a cup. Make a chart to record the temperatures.

4. Which cup had the lowest temperature? Which one had the highest temperature?

____

____

Name ______________________ Date __________

Inquiry Skill: Measure

You **measure** to find out about things around you. You can measure how long, how heavy, or how warm something is.

Learn It

A class wants to measure the temperature in different parts of their classroom. They measure the temperature by a sunny window. They measure the temperature in a shady place. They compare the temperatures after 15 minutes.

a sunny window	75°F
a shady place	70°F

Focus on Skills

Name ______________________ Date __________

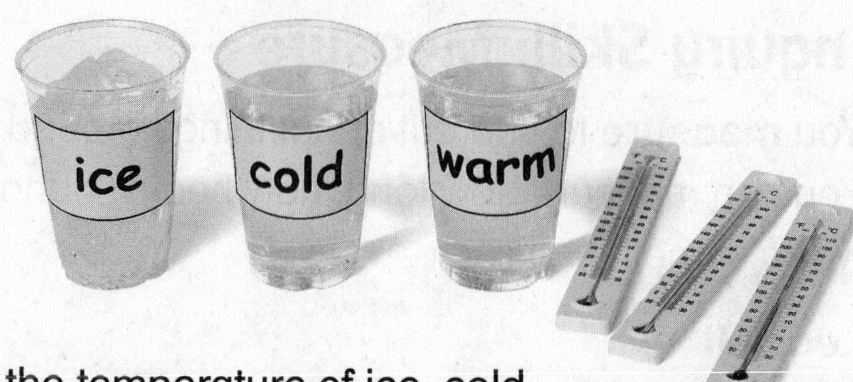

Try It

You can measure the temperature of ice, cold water, and warm water.

1. Fill cups with ice, cold water, and warm water.
2. Predict. What is the temperature in each cup? Record your predictions.

Measuring Temperature			
	ice	cold	warm
predict			
measure			

3. Measure. Put a thermometer in each cup for 5 minutes. Record each temperature.
4. Compare. Were your predictions close to your measurements?

____

____

Name ______________________ Date ____________

Explore

How is sound made?

What to Do

1. Tie the string to the paper clip. Make a hole in the bottom of the cup.
2. Pull the string through the hole. The clip keeps the string from pulling through the cup.
3. Wear goggles. Hold the cup and string with one partner. The third partner snaps the string.
4. **Observe.** What happens? How did you make sound?

____

____

____

You need

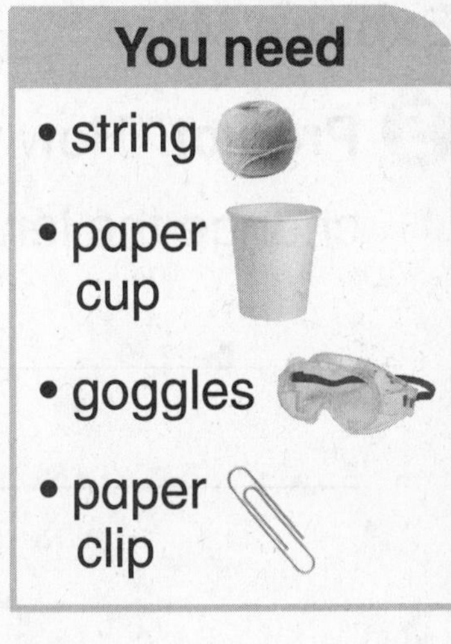

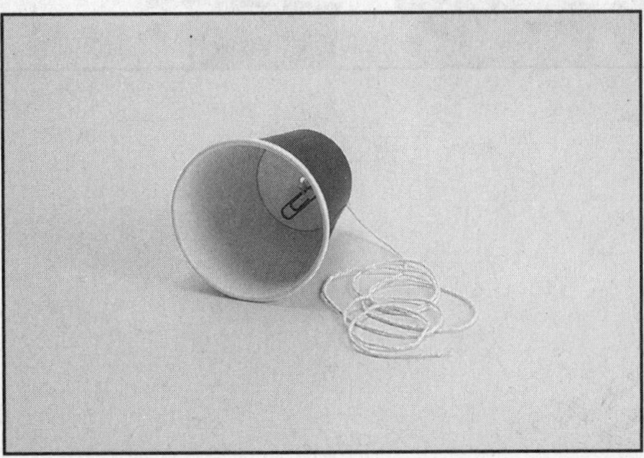

 Name ______________________ Date __________

Explore More

5 **Predict.** How will the sound be different if you change the length of the string? Try it.

____

____

____

Name ____________________ Date __________

Alternative Explore

How can a flute be made out of a straw?

You need

- plastic straw
- scissors

In this activity, you will make a musical instrument out of a straw.

What to Do

1. Make a point at one end of the straw by making two small cuts. Flatten the cut end of the straw by pulling it between your fingers.

2. Describe what happens when you blow into the cut end of the straw.

____

____

____

What Did You Find Out?

3. What caused the straw to make a sound?

____

____

____

Quick Lab

Name ______________________ Date __________

Use a tuning fork to study sound

You need

- tuning fork
- cups
- water

1. Strike a tuning fork against a hard surface. Then hold it in your hand. Why did the sound stop?

____

____

2. **Predict.** Fill a cup with water. What do you think will happen if you strike the tuning fork and hold it in the water?

____

3. **Observe.** Strike the tuning fork against a hard surface. Then hold it in the cup of water right away. What do you see and hear?

____

____

4. What did the water help you see about the way a tuning fork works?

____

____

Name ______________________ Date __________

Explore

What does light pass through?

You need

- flashlight
- cardboard
- plastic wrap
- various items

What to Do

1. **Predict.** Which materials will light pass through? Which will block the light?

____

____

____

2. Work with a partner. Hold up the cardboard. Hold plastic wrap three inches in front of the board. Your partner shines the flashlight on the object.

3. **Observe.** Did the plastic wrap block the light or did the light pass through it?

____

____

4. **Compare.** Which objects block the light and which let light pass through?

____

____

 Name ______________________ Date __________

Explore More

5 **Predict.** What might happen with other classroom items? Try it.

____

____

____

Name ______________________ Date ____________

Alternative Explore

How much light passes through paper?

You need

- flashlight
- different kinds of paper

In this activity, you will explore how light can pass through different kinds of paper.

What to Do

1 **Predict.** How much light will pass through different colors of paper?

____

____

____

2 Hold a piece of paper in front of a blank board. Shine the flashlight on the paper. Try it again with different kinds of paper.

What Did You Find Out?

3 What kinds of paper allowed the most light to go through? Which kinds allowed the least?

____

____

____

Quick Lab

Name ______________________ Date __________

Use a prism to make a rainbow

You need

- prism
- colored pencils
- drawing paper

1 Go outside on a sunny day. How does the sky look? Is there a rainbow?

____

____

____

2 **Observe.** Hold the prism up to the sunlight so that the light shines through the prism and onto a piece of white paper.

____

____

3 **Record.** Use colored pencils to draw what you saw.

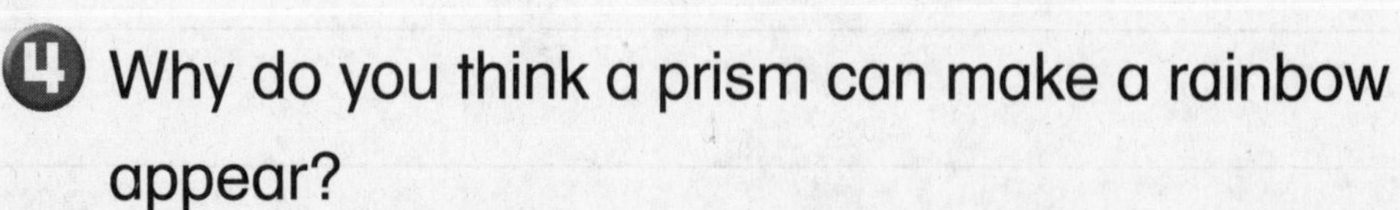

4 Why do you think a prism can make a rainbow appear?

____

____

Name ______________________ Date __________

Be a Scientist

How does sunlight affect the temperature of light and dark objects?

What to Do

1. Record the temperature of each thermometer on a chart. Wrap one thermometer in black cloth as shown. Wrap the other in white cloth.

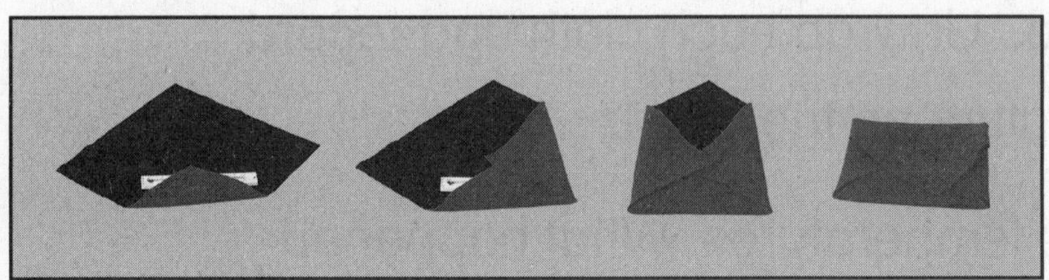

2. Place the wrapped thermometers on a sunny windowsill. Wait 15 minutes.

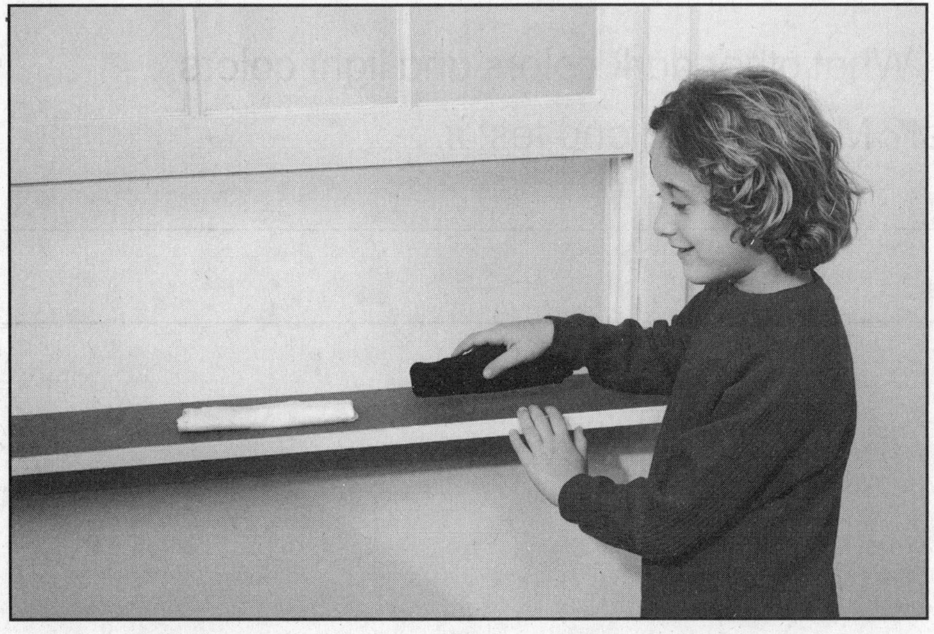

Name ____________________ Date __________

3. **Compare.** Feel each cloth with your hands after 15 minutes. Which color cloth feels warmer?

____

____

4. **Predict.** Which color will have the higher temperature? Why do you think so?

5. **Record Data.** Unwrap each cloth and record each temperature on the chart.

6. Compare the temperatures. What happened to the temperature of each cloth? Was your prediction correct?

Investigate More

Compare. What other dark colors and light colors can you test? Make a plan and test it.

____

____

Name ______________________ Date __________

What makes the bulb light up?

You need

- wires 
- battery
- light bulb 

What to Do

1. **Predict.** Look at the battery, bulb, and wires. How could you put them together to light the bulb? Record your ideas with a partner.

____

____

____

2. **Be Careful.** Try your ideas. Which of your ideas made the bulb light? Which ideas did not work?

____

____

____

3. **Record Data.** Write down your results with your partner. How many ways did you make the bulb light up?

____

____

 Name ______________________ Date __________

Explore More

4 **Predict.** How could you make a second bulb light up? What else would you need?

____

____

____

Name ______________________ Date __________

What does a light bulb need?

You need

- pencil
- drawing paper

What to Do

1. **Observe.** Work with a partner to draw the path of the electric current that travels from the battery to the bulb and back again.

2. **Predict.** What do you think will happen if the electric current cannot make a circuit between the light bulb and the battery?

____

____

3. Draw a diagram that shows an incomplete circuit.

What Did You Find Out?

4. What can keep a light bulb from lighting up?

____

____

Quick Lab

Name ______________________ Date ____________

Make paper come alive

You need
• tissue paper
• markers
• scissors
• plastic ruler
• wool cloth

1. Draw a worm on a piece of tissue paper. Use markers to color it. Then use scissors to cut it out.

2. **Observe**. Put the worm on a flat surface. Rub a plastic ruler on a wool cloth for several seconds. Then bring it close to the worm. What happens?

____

____

3. **Record**. Why do you think this happened? Explain.

____

____

____

____

____

Why Some Fruits Have Many Seeds

You need

- melon seeds
- soil
- measuring cup
- pot
- plastic spoon

Many fruits and vegetables have seeds. Some fruits, like peaches, and plums, have only one seed. Other fruits have hundreds of seeds!

Purpose

Find out why some fruits have many seeds.

Make a Prediction

What might happen after you plant many seeds from a melon?

____

____

Test Your Prediction

1. **Measure.** Fill a pot close to the top with soil.
2. Plant 5 melon seeds. Bury each seed 1 inch below the soil. Water your seeds and put the pot in a sunny place.

Step 2

Name ______________________ Date ____________

3 Record how your seeds grow over the course of three weeks.

Week	Seed Growth

Draw Conclusions

4 Why do you think some fruits have many seeds?

____

____

____

Critical Thinking

5 How do plants keep their seeds safe?

____

6 Why do you think some animals lay many eggs?

____

____

____

Use with Activity Flipchart p. 61

Name ______________________ Date ____________

How Color Helps Animals Hide

Many animals blend into their environment to stay safe. Some toads are brown so they can hide in dirt and mud. The wings of some moths look just like tree bark.

You need

- white paper
- 20 white circles
- 20 brown circles
- stopwatch

Purpose

Find out why some animals grow different color fur or feathers in the winter.

Make a Prediction

What color fur would be hard to see in a snowy place?

____

____

Test Your Prediction

1. Fold the white paper in half. Spread out the circles on one half of the paper.
2. Fold over the other half of the paper to hide the circles.
3. You have ten seconds to pick up as many circles as possible after your partner unfolds the page. Use a stopwatch.

Use with Activity Flipchart p. 62

Name ______________________ Date __________

4 **Record Data.** How many circles of each color did you pick up? Switch roles and try it again.

____

Draw Conclusions

5 How did color help you pick up circles?

____

____

6 **Predict.** What would happen if you did the activity on brown paper?

____

____

Critical Thinking

7 Why do you think many desert animals are brown?

____

____

8 How could you find an animal that blends into its environment?

____

Use with Activity Flipchart p. 62

Name ________________________ Date ____________

Everyday Science

Soil and Sand

Some soils have no sand in them at all. Others are almost all sand. Sand can hold water because there is space between the grains. Topsoil can hold water because it has bits of dead plants and animals.

You need

- sand
- topsoil
- measuring cup
- 2 cups
- tablespoon
- water

Purpose

Find out which dries first, topsoil or sand.

Make a Prediction

How fast do you think topsoil and sand will dry? Which one will dry faster?

____

____

Test Your Prediction

1. **Measure.** Pour one cup of topsoil into a cup. Then pour one cup of sand into another cup.
2. Place both cups in a sunny place.
3. Add three tablespoons of water to each cup.
4. Touch the top of the topsoil and the sand after a few hours.

Name ______________________ Date __________

Draw Conclusions

5 Which stayed more damp, the topsoil or the sand?

____

____

6 Why do you think most plants grow better in soil than in sand?

____

____

Critical Thinking

7 Wind can blow sand away easily. How do you think plants stay in the sand?

____

____

8 Why do you think soils are different colors?

____

____

Name ______________________________ Date __________

Wind Power

You need
• construction paper
• 2 toy cars
• tape
• ruler
• scissors

You already know that the wind can make things move. Sailboats use wind to move across the water. Many plants use the wind to move their seeds to new places. We also use wind to give us energy. When windmills turn, they make electricity. We can use this electricity to heat and power homes.

Purpose

Find out how we use wind.

Make a Prediction

Could a sail help a car move faster?

____

____

Test Your Prediction

1. Make a sail out of paper. Cut out a triangle and then fold it into two equal halves.
2. Tape your sail to one of the toy cars.
3. **Measure.** Make a starting line, and make a finish line that is 20 centimeters away.

Name ______________________ Date __________

4 Place both cars at the starting line. Work with a partner and blow on the cars.

Draw Conclusions

5 Which car crossed the finish line first? Why?

____

____

6 What do you think would happen if the sail were bigger?

____

Critical Thinking

7 Why is wind power useful? What is another way to use wind?

____

____

____

8 Which do you think is a better natural resource, wind or coal? Why?

____

____

Name ______________________ Date __________

Everyday Science

Spin an Egg

You need

- hard-boiled egg
- uncooked egg

Isaac Newton was a famous scientist who came up with the three important laws of motion. The first law of motion says that an object that is not moving will not move until something makes it move. This first law also says that if an object is already moving, it will keep moving until a force speeds up or slows down the object.

It is easy to tell the difference between a hard-boiled egg and an uncooked egg by using Newton's first law of motion.

Purpose

Find out which stops spinning first, a hard-boiled egg or an uncooked egg.

Make a Prediction

Which type of egg will stop spinning first, a hard-boiled egg or an uncooked egg?

____

____

Test Your Prediction

Step 1

1. Spin a hard-boiled egg.

Name ____________________ Date __________

2 While the egg is spinning, grab it with your hand and then quickly let go of it. Observe what happens.

3 Repeat step 2 with an uncooked egg.

Draw Conclusions

4 Which egg stopped spinning first? Why?

____

____

____

Critical Thinking

5 How would you stop a soccer ball from moving?

____

6 Why do you think a ball will not stop moving in mid-air?

____

____

Name ______________________ Date ____________

Everyday Science

The Force of Gravity

You need
- penny
- sheet of paper

The famous scientist Isaac Newton discovered the force of gravity in 1687. As the story goes, he was sitting under an apple tree one day when an apple fell out of the tree and hit him on the head. He wondered why the apple fell down and not up. He explained this as the force of gravity. Even though you can not see it, gravity pulls all things toward the center of Earth. Some things fall faster than others.

Purpose

Find out what happens when a penny and a sheet of paper are dropped at the same time.

Make a Prediction

Will the paper and the penny land on the floor at the same time?

____

____

Test Your Prediction

1. Hold a sheet of paper in one hand and a penny in the other.

Name ____________________ Date __________

2 Hold each item at the same height and drop them at the same time.

3 **Observe.** Watch carefully as each item falls to the ground.

Draw Conclusions

4 Which fell to the ground faster, the penny or the sheet of paper? Why?

____

____

Critical Thinking

5 Which do you think would fall faster, a feather or a pencil? Why?

____

____

6 How could you find out if heavier things fall faster than lighter things?

____

____

____

Name ______________________ Date __________

Move With Magnets

You need

- toy car
- 2 bar magnets
- tape

Every magnet has two poles. Poles are the places on a magnet where the magnet's pull is strongest. Every magnet has a south pole and a north pole. The poles are at opposite ends of the magnet. When two of the same poles get close to each other, they will repel, or push away, from each other. When two opposite poles come together, they will attract, or pull toward each other. Two north poles will repel each other. A north and a south pole will attract each other.

Purpose

Find out how to move a toy car without touching it.

Make a Prediction

What will happen if you put a magnet next to a toy car with a magnet on it?

____

____

Test Your Prediction

1. Tape a bar magnet to the top of a toy car.

Name ______________________ Date __________

2 Use a second bar magnet to push the car.

Draw Conclusions

3 How can you move a toy car without touching it?

____

____

4 Which poles did you put near each other to pull the car?

____

____

Critical Thinking

5 What else do magnets attract?

____

____

6 How can you use magnets?

____

____

____

Name ____________________ Date ____________ **Learning Lab**

What is the best way to grow corn?

You need

- half an ear of corn
- pan
- water

Ask Questions

What does a seed need in order to grow? How much water does a seed need to grow? Does a seed need soil to grow?

Make a Prediction

Will corn seeds grow with different amounts of water?

____

____

Test Your Prediction

1. Place half an ear of corn in a pan. Lay the corn on its side.
2. Pour water in the pan until half of the ear is underwater.

Name ______________________ Date __________

3 **Predict.** How many seeds will sprout? Where will they sprout?

____

____

4 Change the water every two days. Make sure you keep the water at the same level. Do not let the corn roll over.

5 **Observe.** Watch your ear of corn grow for two weeks. Record how your corn changes.

Week	Changes
1	
2	

Communicate Your Results

What happened to your corn? Discuss your results with a partner.

▶ How did your predictions compare with your results?

____

____

▶ Can seeds have too much or too little water to grow?

____

Use with **Activity Flipchart** pp. 68–69

Name ______________________ Date ____________ Learning Lab

Reach for the Sky

You need

- 6 corn seeds
- clear cup
- soil
- water

Ask Questions

Will corn seeds grow if they are planted in soil? Will they grow if they are planted close to the surface?

Make a Prediction

What will happen if you plant corn seeds under too much soil?

____

____

Test Your Prediction

1. Put two corn seeds in a clear cup. Put the seeds against the side of the cup so you can see them. Cover the seeds with just a little soil.

Name ____________________ Date __________

2 Put two corn seeds against the side of another cup. Add a lot of soil on top of the seeds.

3 Put two more seeds in a third cup. Add soil until the seeds are 1 inch below the surface.

4 **Predict.** Which seeds will sprout?

____

____

Name ______________________ Date ____________

Learning Lab

5 **Record Data.** Keep the soil moist, and record how your seeds change.

Week	Changes
1	
2	

Communicate Your Results

Discuss as a group what happened to your seeds.

- ▶ Which seeds sprouted?

____

____

- ▶ Which seeds grew fastest?

____

____

- ▶ Which seeds started and then stopped sprouting?

____

____

Name ______________________ Date __________

Corny Experiments

What helps plants grow? What keeps plants from growing? Answer the questions below.

▶ Can corn grow in sand? Can it grow in clay?

____

____

____

____

____

▶ Does temperature change how seeds sprout? Would your seeds sprout in the refrigerator?

____

____

____

____

____

Name ______________________ Date ____________

Learning Lab

How do we use natural resources?

You need
- soil
- plate
- hand lens

Ask Questions

Most plants need soil to grow. What is in soil? How many different things can you find in soil? Are all soils alike?

Make a Prediction

What do you think you will see in soil if you use a hand lens?

____

Test Your Prediction

1. Place some soil on your plate.
2. Use a hand lens to observe your soil.

Name ______________________ Date ____________

3 **Classify.** Find objects that look alike. Identify them as plants, animals, or rocks.

4 **Record Data.** Make a chart. On your chart, write and draw the objects you found.

Plants	Animals	Rocks

Name ______________________ Date __________

Communicate Your Results

What did you find in the soil? Discuss your results with a partner.

▶ How did your predictions compare with your results?

___

___

___

▶ What did your partner find in his or her soil?

___

___

___

▶ How do you think the objects in the soil got there?

___

___

___

Learning Lab

Name ______________________ Date ____________

Fun With Cotton

You need

- cotton plant
- hand lens

Ask Questions

We eat many kinds of plants, but we also use them for other things. How do we use plants? What can we make out of cotton plants?

Make a Prediction

Why do people grow and pick cotton? Write a prediction.

____

____

Test Your Prediction

1. **Observe.** Use a hand lens to look at a cotton plant. What do you see?

____

____

2. Carefully take the fluffy, white ball off of the plant. Take out all of the seeds.

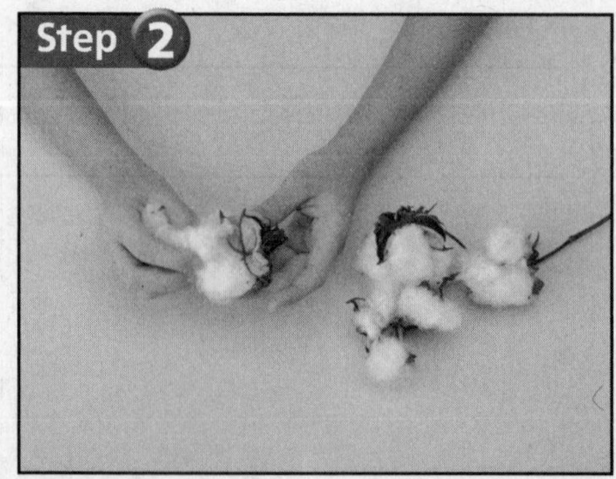

3 Communicate. What does the white part feel like?

____

4 Gently pull the cotton apart with your fingers. Stretch and twist the cotton to make thread.

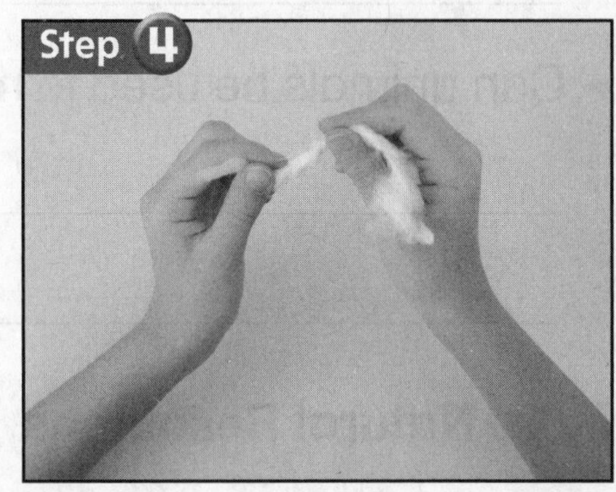

Communicate Your Results

Discuss your results with a partner.

▶ What are some uses for thread?

____

____

____

 Name ______________ Date __________

▶ What items that come from plants do you use every day?

▶ Can animals be used to make thread?

More Natural Resources

We use many things from nature. Answer the questions below:

▶ How can the Sun be used to heat water?

▶ How can wind be used to cool water?

▶ How can water be used to move things?

Use with Activity Flipchart pp. 70–71

Name ____________________ Date ____________

Learning Lab

How can we test a magnet's strength?

You need
- 3 magnets
- ruler
- white paper

Ask Questions

What happens when magnets are next to each other? How far apart can two magnets be and still attract each other?

Make a Prediction

How will adding more magnets affect the strength of the attraction?

____

____

Test Your Prediction

1. Place a magnet on a piece of paper. Trace the magnet.
2. Place a second magnet on the paper. Move it toward the first magnet until the first one moves. Make another mark to show where the second magnet was when the first one moved.

Name ______________________ Date ____________

3 **Measure.** How far apart were the magnets when they attracted each other?

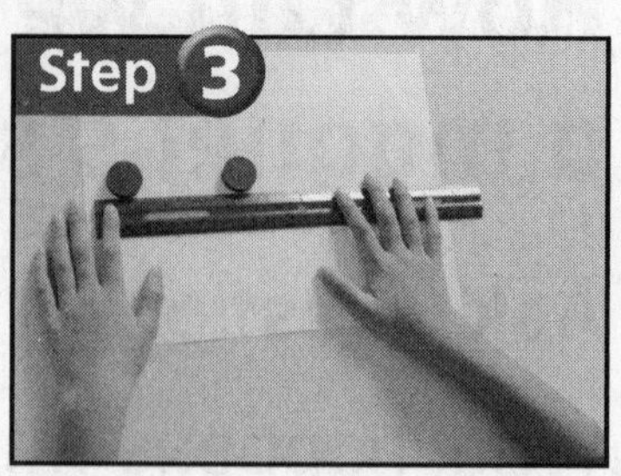

____

____

4 Now use two magnets to move the first magnet. How far apart were the magnets when they first attracted each other?

____

____

Communicate Your Results

Discuss your results with a partner.

▶ How did your prediction compare to your results?

____

____

▶ Are two magnets stronger than one magnet? How do you know?

____

____

Name ______________________ Date __________

Pulling Through Water

You need

- magnet
- paper clips
- clear plastic cup

Ask Questions

Magnets attract objects that are made of certain types of metal. Magnets can also attract objects through some types of materials. What can magnets pull through?

Make a Prediction

Can a magnet attract paper clips through a cup of water?

____

____

Test Your Prediction

1. Put some paper clips in a clear plastic cup. Hold one end of the magnet against the side of the cup.

Name ______________________ Date __________

2 How many paper clips did the magnet attract?

____

3 Fill the cup with water and repeat the activity.

Name ______________________ Date ____________ **Learning Lab**

4 Record Data. Fill in the chart to share your results.

	Number of Paper Clips
Cup	
Cup with water	

Communicate Your Results

Discuss your results with a partner.

▶ When did the magnet attract more paper clips?

____

____

▶ What can magnets pull through?

____

____

Learning Lab

Name ______________________ Date __________

Strength Test

What else can magnets do? Answer the questions below:

▶ Where on a magnet is the pull strongest?

____

▶ How many pieces of paper can a magnet pull through?

____

____